GLOBAL CHANGE AND CHALLENGE

Revolutionary changes to our world economy and their impact on the environment are presenting formidable problems for human society. *Global Change and Challenge* examines some of the crucial challenges facing society in the 1990s and how geography can contribute to their understanding and management.

Using the theme of how societies adapt to change, the contributors – at the time of writing all members of the Geography department of the London School of Economics – deal with some of the issues confronting modern geography in three broad groups. The early chapters examine the impact of human activity on the environment in a global context – looking in particular at the management of resources and of the natural environment. The book then examines from a variety of perspectives how global economic change offers both new constraints and new opportunities for economic development at all levels – local, regional and global. The shifting patterns of social and economic change are examined with respect to the industrialised, newly industrialised, and Third World countries at both national and regional levels. The final chapter assesses how new technology can influence the geographer's perspective and aid in the search for management solutions.

By employing the general theme of adaptation to change, the contributors seek to present a range of views on the 'geography of change' in a single and accessible form for school and university students. The aim is as much to encourage students to understand where we are and where we have come from, as to where we may be going.

Robert Bennett is Professor of Geography at the LSE, having previously held positions at Cambridge University and UCL. His recent publications include *Decentralisation, Local Governments and Markets* (Editor, Clarendon Press, 1990) and *Local Economic Development* (Bellhaven, 1991). **Robert Estall** is Emeritus Professor of Geography at the LSE, from where he graduated in 1955. He has had three extended stays in the United States as Visiting Professor or Research Fellow, but his academic career has centred on the LSE. His major publications include *Industrial Activity and Economic Geography* (Hutchinson, 1980), *New England: A Study in Industrial Adjustment* (Bell, 1966) and *A Modern Geography of the United States* (Pelican, 1976).

GLOBAL CHANGE AND CHALLENGE

Geography for the 1990s

Edited by
Robert Bennett and Robert Estall

DEPARTMENT OF GEOGRAPHY
LONDON SCHOOL OF ECONOMICS

London and New York

First published 1991 by Routledge
11 New Fetter Lane, London EC4P 4EE

Simultaneously published in the USA and Canada
by Routledge
29 West 35th Street, New York, NY 10001

Reprinted with corrections 1994

Transferred to Digital Printing 2004

© 1991 R. J. Bennett and R. C. Estall

Typeset by J&L Composition Ltd, Filey, North Yorkshire

British Library Cataloguing in Publication Data
Global Change and Challenge: Geography for the 1990s
1 Geography
I Bennett, R. J. (Robert John) II Estall, R. C. (Robert
Charles)
910

Library of Congress Cataloging in Publication Data
Global Change and Challenge: Geography for the 1990s/Edited
by Robert Bennett and Robert Estall.
p.   cm.
Includes bibliographical references and index.
1. Geography   I. Bennett, R. J. (Robert John)   II. Estall,
Robert C
G128.G57 1991
910—dc20   90–27303

ISBN 0–415–00142–0
ISBN 0–415–00143–9 pbk

CONTENTS

CONTENTS

PLATES

FIGURES

FIGURES

TABLES

CONTRIBUTORS

At the time of writing, the editors and contributors to this book were all members of the Geography department of the London School of Economics and Political Science, Houghton Street, London WC2A 2AE.

Robert Bennett is Professor of Geography, and teaches on the economic and urban geography of Britain and Europe.

Christopher Board, Senior Lecturer, is Britain's leading university cartographer and teaches cartography and geographical methods.

Sylvia Chant, Lecturer, teaches social geography of the Third World emphasizing employment, urbanization and gender.

Derek Diamond is Professor of Geography and Urban and Regional Planning, and teaches on urban change, inner cities and planning.

Robert Estall is Emeritus Professor of Geography of North America, and is a specialist in economic geography.

Ian Hamilton, Senior Lecturer, teaches on global industrial change with special reference to the European Community, Eastern Europe, the Soviet Union and NICs.

Michael Hebbert, Senior Lecturer, teaches on urban development and planning policy.

David Jones is Professor of Physical Geography, and teaches geomorphology and environmental management.

Judith Rees, previously Senior Lecturer at the LSE and now Professor of Geography at Hull University, is an expert in resource economics.

Helen Scoging, Lecturer, teaches erosion and environmental systems in semi-arid environments.

INTRODUCTION
Robert Bennett and Robert Estall

PERSPECTIVES ON CHANGE

This book examines some of the major current challenges of our society and how geography can contribute to their understanding and management. It presents a diversity of views by authors working in one university geography department at the London School of Economics.[1] It does not seek to provide the answer to all questions, nor to give an exhaustive coverage of all issues of concern to modern geographers. Instead, one theme is examined: the adaptation of society to change. The aim is to present a range of views on the theme 'geography of change' in a single and accessible form for school and university student readers. We also suggest ways in which geography might help to arrive at management solutions to major current policy problems. This is part of the tradition of the discipline. It has been the particular role played by the LSE department since at least the 1930s when, under Professor Sir L. D. Stamp, it directed the Land Utilisation Survey of Britain.

The issues confronting modern geography are dealt with in three broad groups. The early chapters of the book seek to put the study of change and development in a global context. The problems experienced in British society are not unique, but are part of a more general restructuring of the economies and environments of the world. In the first three chapters we examine the impact of environmental change. The problems that we are now facing are the result of economic and technological developments that are placing increasing pressures on the physical world. In Chapter 1 Judith

---

[1] One author, Judith Rees, has now moved to become Professor of Geography at Hull University, but her teaching interests have been enlarged at the LSE with new staff appointments.

1

Rees examines the great challenges to the management of our environment and its resources, particularly in the fields of minerals and water use. David Jones then reminds us, in Chapter 2, that in spite of our growing ability to manage and regulate the environment, the problem of natural hazards and disasters, such as floods, hurricanes and drought, often on a massive scale, are ever present. One major set of hazards and environmental management pressures is then examined in more depth. In Chapter 3 Helen Scoging demonstrates how desertification is a problem arising from the way in which different societies use, and abuse, the environment.

In the next seven chapters we examine how global economic change offers both new constraints and new opportunities for economic development. Such changes place regions and towns in new positions relative to each other, relative to the governments of their countries and relative to the rest of the world. These chapters give particular attention to the transformation of the world economy through the institutions of multinational and transnational corporations, as well as via modern production technology and the revolution in communications.

Ian Hamilton, in Chapter 4, shows how global change and restructuring are leading to a new role for the 'Pacific Rim' countries, especially for Japan. A shift of the 'core' of world economic activity is taking place, producing major new 'control' centres. This, in turn, creates particular difficulties for the older industrial centres. Robert Bennett, in Chapter 5, takes up this theme at the national level and examines developments in capital investment and labour skills which have led to contrasted experiences of regional growth, decline and revitalisation. The problems of adapting to change within 'old' regions are examined by Robert Estall in Chapter 6, using the particular example of the USA. This theme of regional change is further developed in Chapter 7, where Robert Bennett and Robert Estall examine a range of case studies of economic development – for the 'old' region of New England, for the new 'silicon landscapes' in the USA and Britain, the Japanese 'technopoles', 'Third Italy' and worker co-operatives.

The Third World is another major arena of change, containing over three-quarters of the world's population. The way in which the Third World develops economically and socially has enormous significance for us all. These developments are examined by Sylvia Chant for the Third World as a whole in Chapter 8 and for the internal geography of Third World countries in Chapter 9.

In the next two chapters we again change the focus to consider certain more local problems of change that are faced in the advanced economies. In Chapter 10 Michael Hebbert examines the tensions that exist between town and country, and the role of the planning system in managing them; while, in Chapter 11, Derek Diamond examines the management problems created by urban change, illustrating them chiefly from British experience.

Finally, in Chapter 12 Christopher Board describes how the use of new means of communication can affect the geographer's perception and solution of problems. The geographer's role is to offer means to develop and present information in new ways by using innovative forms of computer information systems, and devising new uses for traditional maps and diagrams. The development of these new presentations of data, using geographical information systems, can itself be fruitful in suggesting new management solutions.

OUR PLACE IN TIME

The book seeks to look forward. We are now experiencing a period of unprecedented change as part of what is a revolution of our world economy and society and a growing impact on the environment. We ask where the future will lie. Study of the present position allows us to see how our roles are changing. But it is clear that many further changes will be needed everywhere in our global community if our societies are to be sustained.

One aspect of change is the use of education to found a set of lifetime skills that will help students and teachers alike to adapt learning to new demands. Indeed, the challenge of change is as much a problem of finding ways of adapting old and familiar practices as it is of understanding where we are going in the future. The management of change therefore requires us to understand where we are, and where we have come from, as much as where we may be going.

The individual chapters of the book are each intended as statements of this belief. They are offered as case readings to stimulate the student to understand each geographical problem, the issues that arise in its management, and how progress can be achieved. To aid learning, each chapter is accompanied by a brief list of further reading and by a range of essay or discussion questions, drawing on the material covered in each chapter.

1

RESOURCES AND THE ENVIRONMENT
Scarcity and sustainability
Judith Rees[1]

Change in the economy and society of the countries of the world has led inevitably to changes in the role of resources and the environment. Resources are the keystone of civilisation. The development of individual nations and, indeed, of the entire international economic and social system, is dependent on their continued availability. Given their importance, it is understandable that fears of impending resource scarcity have periodically emerged. Among the most well known are, perhaps, the gloomy forecasts of Malthus in the late eighteenth century, and the recent studies of 'the limits to growth'. Such fears have been based on the belief that either the physical system, or human institutions, or both, must fail in the relatively near future to deliver the extra resources needed to satisfy growing human demands. These fears provide the focus for this chapter.

Traditionally, concern has been centred on the limited capacity of the world to provide the minerals essential for economic development. However, it has long been clear that the physical adequacy of mineral stocks is in many ways the least critical part of the problem. Apart from physical scarcity, other scarcity problems exist that can be grouped together as 'geopolitical', 'economic' annd 'environmental' (Table 1.1). Recognition that scarcity is not one problem, but many, is crucial not only to our understanding of the issues but also to the development of resource management policies. The reality of threats of scarcity varies between problems and also over time and geographical space. In addition, given the range of concerns about scarcity, it is now clear that there is no one set of

---

[1] Judith Rees is now Professor of Geography, University of Hull.

Table 1.1 The dimensions of scarcity

Type of scarcity		Concern
Physical scarcity	1	Exhaustion of minerals and energy;
	2	Human populations exceed the food production capacity of the land;
	3	Depletion of renewable resources such as fish, soils or timber.
Geopolitical scarcity	1	Use of mineral exports as a political weapon (e.g. sales embargoes);
	2	Shift in location of low cost mineral sources to 'hostile' blocs of nations.
Economic scarcity	1	Demand at current price levels exceeds the quantity supplied (therefore shortages);
	2	Needs exceed the ability of individuals or countries to pay for resource supplies;
	3	Rich economies can always outbid the poor for essential resources, creating unequal patterns of resource use;
	4	Economic exhaustion of specific minerals or renewable resources causes economic and social disruption in producer regions or in nations dependent on them.
Renewable and environmental resource scarcity	1	Disruption of essential biogeographical cycles (e.g. the carbon dioxide cycle and the greenhouse effect) threatening sustainability of life on earth;
	2	Pollution loads exceeding the 'absorptive' capacity of the environment causing economic, health and amenity problems;
	3	Loss of plant and animal species and landscape values, with wide, but poorly understood, long-term consequences.

causes common to them all, nor is there a single appropriate policy response.

THE NATURE OF RESOURCES

We must understand the term 'resources' and classify them before we can discuss the various problems arising from potential scarcity and evaluate the threats they pose.

Natural resources are those products or properties of the physical environment which human beings are technically capable of utilising and which provide desired goods and services. Both these criteria must be satisfied before a particular part of the physical world can acquire a value as a resource. Technological innovation and improved knowledge only create the opportunities for utilisation. Whether these opportunities are taken up depends upon economic, social and political demands. Resources are, therefore, defined by human desires, needs and capacity. They are phenomena which depend greatly on the prevailing culture. This helps to explain the many disagreements over which particular elements in the environment are resources.

Just as it is difficult to define what resources are, so the task of assessing whether future supplies will be adequate is also highly complex. We have not only to evaluate the capacity of the world to sustain adequate resource supplies over time, but we must also assess the extent to which human intervention (intended or unintended) will act to change those supplies. In addition we must try to predict which particular resources will be judged to have value by future generations.

As human societies have evolved so has the definition of potentially usable resources. Today's list vastly exceeds that of Stone Age economies. Inevitably, too, reappraisal will continue in the future. Innovations and changes in life style will continue to change natural substances or environmental properties from 'neutral stuff' into valued resources. It must be stressed that this transfer process is not one-way. Just as flint lost value when metal tools were developed, so copper, coal, tin or uranium could all lose value, or revert to neutral stuff, if more effective or lower cost substitutes are found, or if consumer tastes change. It may be, for example, that the risks of radiation could eventually become so politically unacceptable that nuclear power generation will be phased out and uranium will lose the resource value it acquired in the 1940s, when atomic energy was first harnessed.

Since the value of resources is determined by the culture that uses them, it follows that the value of particular resources varies not only over time but also over space. But the geographical diversity of the value given to metals and fuels has been reduced by modern communication systems and expanding world trade. Generally the global value of such resources is determined by the demands and technologies of the advanced nations. However, there is much less international consensus over the assessment of environmental resources such as air, landscapes, wilderness areas or plant species. To a Brazilian peasant farmer, for example, the tropical rain forest may simply be an impediment which must be removed before the valued resource, land, can be utilised. The notion that the forest itself is a vital resource, either through its contribution to the global carbon cycle or because of the diversity of tropical forest species, is unlikely to mean much to the farmer. Similarly, the long term possibilities of global warming of the climate are of little concern to those living on the edge of starvation.

Conflicting assessments of environmental resources can exist even among individuals sharing a common cultural heritage and living in the same small community. What for a local farmer is a weed-strewn piece of unproductive wasteland could be a rich aesthetic and ecological resource to others. Such differences in valuations and priorities lie at the heart of many of the current conflicts over environmental protection.

THE RESOURCE CONTINUUM

It has been conventional to split resources into two types – non-renewable (stocks) and renewable (flows). This terminology is misleading, however, and it has limited value in the debate over future resource availability. All resources are renewable on some time scale – new oil, coal, natural gas and metal deposits are being formed today. What matters for the sustainability of future supplies is the relative rates of replenishment and use. Moreover, replenishment cannot simply be regarded as a natural process. In some cases, human intervention can significantly alter the rate of replenishment. It seems better, then, to think in terms of a resource 'continuum' rather than of stocks and flows (Figure 1.1).

At one end of the continuum are the fossil fuels, such as coal and oil, which are being formed too slowly for the replenishment process to be relevant to human needs. They are consumed by use and cannot, as yet, be replenished artificially (at least, not at reasonable price levels and in significant quantities). Heavy use of such resources cannot be sustained indefinitely and scarcity will ultimately occur unless substitutes are found. At the other end of the spectrum is a group of resources which should be infinitely renewable and the supplies of which are unrelated to current use levels. This group includes solar, tidal and wind energies and the global system of water circulation. However, the word 'should' needs to be emphasised. Evidence is accumulating which suggests that inadvertent human intervention can affect levels of incoming and outgoing solar radiation and, in so doing, can also alter

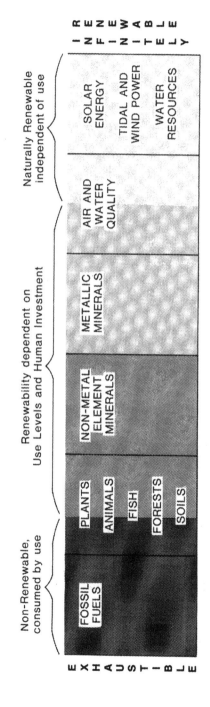

Figure 1.1 The resource continuum

precipitation patterns. If this is the case, then such resource flows will no longer be naturally but partly affected by human activity.

Most resources lie between the extremes of the resource continuum. Hence their future availability depends to a large degree on human choices. The supply of those resources that reproduce biologically, for example, can be sustained only by ensuring that use rates remain below natural replenishment rates, or by investing in planting or breeding programmes. Unless the use/investment balance is maintained such biological resources can be 'mined' to extinction. Likewise the continued supply of minerals such as iron, lead or copper, will depend on the willingness and ability of society to invest in recycling. However, unlike the fossil fuels and biological resources, the element minerals cannot be destroyed by use; human beings therefore, retain the option to collect and recycle them, if necessary, in the future. The future availability of acceptable air and water quality will also be determined by the levels of investment in pollution abatement and the development of technologies to increase the natural assimilative capacity of the environment. River water quality, for example, can be improved by adding oxygen, which increases the speed with which natural processes break down pollutants.

Once we are aware of the cultural definition of resources and the nature of the resource continuum, there is nothing inevitable about scarcity. Any future problems of resource scarcity will not be caused by nature but by the failure of human institutions to adopt appropriate management practices and develop adequate substitutes. How likely is it, then, that scarcity problems will arise? To examine that question we return to the four types of scarcity listed in Table 1.1.

PHYSICAL SCARCITY: MINERALS

A great debate raged in the late 1960s and early 1970s over the imminent scarcity of mineral and energy supplies. It was argued that this scarcity would bring about a collapse of the global economic system. Today the debate appears remote, even irrelevant. The pressing problems now being faced, particularly in Third World nations, are of mineral surpluses and falling prices of primary products. However, continuing surpluses are no more inevitable than was imminent scarcity. Future availability depends on three key variables:

1 Investment in searching for and developing new sources.
2 Development of substitutes.
3 Levels of demand.

In an ideal free market system these variables would adjust automatically to prevent both shortages and surpluses. When any mineral is in short supply its price rises. This triggers a whole series of changes in demand, supply and technological effort. These should eventually overcome the shortage (Figure 1.2). Rises in price in the short term lead to falling demand as consumers practise greater economy in use, turn to substitutes or adopt more recycling. If prices remain high for long enough, technological innovations will be stimulated. This enables demand for the resource to fall still further in the longer term. At the same time, changes in supply will also occur. Rising prices will allow the exploitation of known, but

Figure 1.2 The idealised market response to resource scarcity

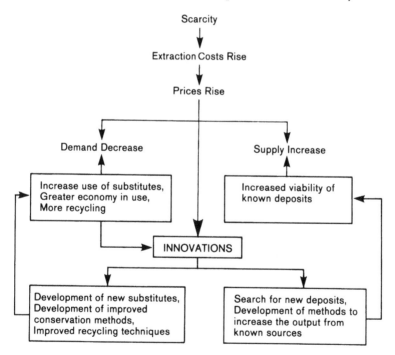

previously uneconomic, sources of supply, promote the development of more efficient extraction technologies and encourage the search for new sources. The same process works in reverse if surpluses result in falling resource prices.

Clearly, such supply responses cannot go on for ever. With rates of use of the fossil fuels, for example, far exceeding the exceptionally slow pace of natural replenishment, absolute physical limits will in time be reached. However, physical exhaustion need not mean scarcity in a functional sense. If alternative ways can be found to provide the same goods and services, then the exhaustion of a particular physical substance need not matter. Optimists thus argue that no one resource is irreplaceable; in time substitution will occur or the same mineral may be found in different geological sources. For example, if bauxite supplies fail, then aluminium could be obtained from carboniferous shales, kaolin clays or other widely available substances. Alternatively, the scarce mineral could be replaced by entirely different materials capable of fulfilling the same functions. For example, solar, nuclear or tidal energy could replace fossil fuels. It is also possible that technology will provide the required substitutes. Modern communication technologies using the silicon chip have already reduced the demand for metals such as copper. For many minerals, too, shortages of freshly mined products could be overcome by increased recycling. Finally, a kind of substitution could occur in a rather different way. Human life styles and patterns of demand may change and alter the mix of resources in demand. According to some analysts such changes are already occurring as people in 'post-industrial' societies shift away from demanding more material goods and purchase more services, which require fewer material resource inputs.

The scope for substitution is immense and the robustness of the market response to threats of scarcity can be seen in the speed with which adjustments were made following the large oil price increases in 1973 and later years. In the case of oil it was geopolitical scarcity, not physical exhaustion, that produced massive price rises (Figure 1.3). The Organisation of Petroleum Exporting Countries (OPEC), formed in 1960 to increase the revenues of oil producing nations, successfully pushed up the price of oil in 1973. But the response mechanism of substitution came into play. As a result the contribution of oil in the global pattern of energy use declined after 1973 as consumers turned to coal, nuclear power and the renewable

12

Figure 1.3   Oil price index, 1974–94

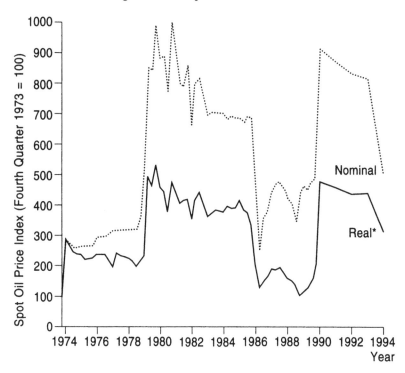

* Adjusted for Inflation and the Dollar's Devaluation

resources. New non-OPEC oil sources were developed, and energy conservation measures were introduced (Figure 1.4). As a result, oil prices fell sharply in the 1980s and OPEC's contribution to world oil production fell from 53 per cent in 1973 to 32 per cent in 1987.

Market changes of this type are bound to impose costs of a socio-economic and environmental nature. But most analysts now agree that the economic system in advanced countries has sufficient mechanisms to allow it to adapt to shortages of specific resources when these are needed to sustain growth and capital accumulation. In other words strong economic self-interest is a powerful force preventing physical scarcity of key resources, at least in advanced nations.

Figure 1.4 The declining oil import dependence of industrial countries, 1970–86

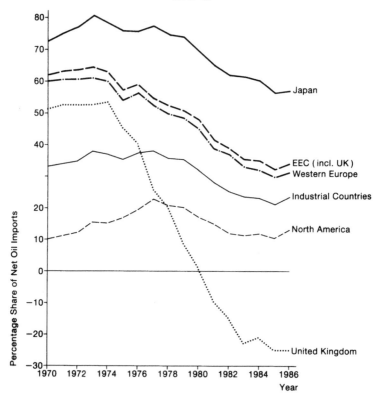

Note: EEC data includes Greece after 1981, and Spain and Portugal after 1986.

GEOPOLITICAL SCARCITY

Concern over the physical scarcity of essential minerals in the 1970s was compounded by the popular notion that Third World and socialist country mineral producers could gain a monopoly over the supply of key resources. Such fears were derived from the observation that advanced Western nations, with their long histories of resource exploitation, had exhausted their own low cost supplies and would, therefore, have to rely increasingly on imports. Import dependence would then leave them vulnerable to price rises, such as those imposed by OPEC, or trade embargoes.

Such shortages are geopolitical in nature. They could conceivably cause short term problems for the capitalist countries, but many analysts now believe that few significant medium to longer term threats are posed. There are several reasons for this optimistic view. First, it is argued that the developed countries, as a whole, have normally taken 60 per cent or more of their energy and non-fuel mineral requirements from other developed countries. This position is unlikely to change in the foreseeable future. Over the last twenty-five years approximately 80 per cent of all investment in the exploration and development of new mineral sources has been concentrated in politically 'safe' advanced countries, particularly Australia, Canada, the United States and South Africa.

Second, a long history of mineral development in a particular region does not necessarily mean that future supplies will be high in cost. The assumption made by the geopolitical scarcity pessimists is that areas with the most favourable geological conditions, or deposits with the highest quality minerals, offer the lowest costs, and will be used up first. But this assumption is based on a misunderstanding of the nature of mineral reserves and of the factors which determine total exploitation costs. When all costs (for infrastructure, mining, milling, transport, administration, taxation and marketing) are taken into account, it is clear that the issues are not that simple. Canada, South Africa and the United States, for example, all possess relatively low grade copper deposits. Nevertheless, their overall mining, transport and marketing costs are below those of the relatively high grade copper ores mined in Zambia, Zaire and Peru.

One crucial cost saving for established mining centres in advanced countries is that the necessary infrastructure is normally available. Rarely has it all to be provided by the mining company. On the other hand, when developing a new source in a Third World

15

JUDITH REES

country, mining companies may have to undertake massive investment in transport facilities, energy and water supply, workers'
housing and so forth. Similarly, the advanced nations have the
capital and technology to help ensure that their mining costs remain
low. Highly mechanised production techniques enable even very
low grade sources to become viable in such countries.

A third reason for suggesting that geopolitical scarcities will not
be a significant problem in the foreseeable future is that, for most
minerals, current production sites, as well as known or potential
reserves, are geographically widespread. They occur in numerous
countries with varied political affiliations and economic interests.
The stimulus to diversify supply sources can have many causes, but
one major factor has been the rise of nationalism in the less
developed countries and the threat this was thought to pose to
multinational companies and consumer nations. The proverb 'Don't
put all your eggs in one basket' has clearly been taken to heart.
Diversifying sources of supply reduces the likelihood of geopolitical
scarcity, since it makes it much more difficult for producers to get
together and take concerted action against consumers.

A fourth consideration is that few minerals have no substitutes. It
is possible, of course, to find cases where there are, as yet, no
satisfactory substitutes for a given mineral in a particular end use.
There is still no satisfactory substitute for the cobalt needed in
carbides for machinery and machine tools, while chromium is
essential to the production of stainless steel. However, these non-
substitutable end uses normally account for a small proportion of
total demand, and it is highly unlikely that all sources of supply of
the mineral would fail. The availability of substitutes, and the
possibilities of recycling, mean that the same economic forces that
work to overcome physical inadequacy can also operate to minimise
the risk of geopolitical scarcity.

Finally, it now seems likely that both the less developed and most
of the old 'socialist' countries are far too dependent on the earnings
from their resource exports to wage a 'resource war'. The need for
foreign exchange, investment capital, plant and machinery and
Western technology ties both groups of countries into interdependence with the West. As the aftermath of the 1970s oil crises clearly
showed, economic recession in the advanced Western economies
soon has major effects elsewhere in the world trade system. This
places severe constraints on the use of resources as political weapons
by the producer nations.

16

All in all, it now appears that earlier fears of physical shortages or geopolitical scarcity were largely unfounded. However, this does not mean that society does not face any scarcity problems. Generalising broadly, there are two main areas where technological progress, market forces and political responses have conspicuously failed to stave off the advent of scarcity. These are in the field of economic scarcity, demonstrated in the supply of the resources needed for the development of many Third World countries, and in the protection of renewable and environmental resource supplies. Each of these is discussed below.

ECONOMIC SCARCITIES

Major problems of resource adequacy are occurring, and will undoubtedly continue to occur, in less developed countries (LDCs), where they affect the well-being of many millions of people. Water shortages are endemic for nearly half the world's population; soil erosion and desertification are creating critical food scarcities, particularly in the semi-arid areas of Africa; the chief domestic energy resource in many LDCs – fuelwood – is in desperately short supply as deforestation continues; and economic development in countries such as Tanzania is crippled by lack of energy supplies and metals. However, such scarcity problems rarely mean that some absolute physical limits to supply have been reached. Rather, the limiting factors are economic and cultural. The big impediments to development are low purchasing power and a lack of sufficient capital, knowledge, technology, skilled manpower and appropriate management institutions.

Water, food, timber and energy have for long been marketable products, with a recognised price, even in the poorest areas of the world. But, unlike in the advanced countries, market mechanisms have not operated to ensure adequate supplies. The reasons for this are complex, but one factor is of fundamental importance. In developed countries the market operates to reduce scarcity by increasing prices, which reduces demand and stimulates investment in new supplies. However, this process is dependent on the price people are able to pay being sufficient to provide the necessary incentives to increase supply. In less developed countries, where households live in abject poverty, price rises tend simply to increase the problems of scarcity. Higher prices mean that people can no longer afford to buy the food, fuel or water they need even to

17

survive at 'baseline' standards of living. In these circumstances, the consumers lack the effective demand to justify private sector investment to overcome the problem of scarcity.

Improvements in supply are made even more difficult if the resource that is marketed competes with a free, naturally available, supply. For example, a local river could provide a substitute, albeit a highly polluted one, for piped water. Gathering fuelwood from open access areas in the countryside could be an alternative to buying wood in the market place. Governments concerned to protect public health or prevent deforestation (and thus limit the attendent problems of soil erosion and desertification) may wish to discourage the use of these alternative sources, and so subsidise piped water and wood production. However, such subsidies cost money and, unless aid from overseas is available, these subsidies divert much-needed resources from other uses. With a limited supply of financial capital a country can provide more water or wood only at the expense of less investment in industry, agriculture or energy. Thus the cycle of underdevelopment and poverty goes on.

Further problems arise when expansion of supply can occur only by importing resource products or by importing the capital and technology needed to develop potential local supplies. It is necessary to remember that import prices are largely determined by the income and demand levels in the advanced countries. Inevitably this means that the supplies available to Third World nations are severely limited by inability to pay. In times of relatively tight world supplies and high prices, available resources are invariably channelled to countries with the greatest buying power. During the 1970s oil crises, for example, the countries hardest hit by price increases were the oil-importing LDCs. Even during periods when particular products are in global surplus, shortages can be acute in low income countries simply because they lack the necessary foreign exchange.

Such foreign exchange shortages also place severe limits on the ability of many countries to develop indigenous resource supplies. This is particularly so today, when Third World debt burdens, accumulated in the late 1970s and early 1980s, mean that a high proportion of export earnings is required simply to service the loans. Often 40–50 per cent of export earnings has to be devoted to debt repayment. This also makes further borrowings out of the question. In these circumstances it is all too easy for countries to fall into a 'development trap'.

Tanzania, for instance, became heavily indebted in the 1970s by importing expensive oil for its industry and transport system. This debt meant that it had to strive even harder to increase exports to pay off interest. But interest payments to service the debt now take such a high proportion of export earnings that Tanzania can no longer afford to import the oil, machinery and transport equipment needed to keep its exports flowing. Likewise, the Zambian economy is rapidly falling into chaos because its foreign exchange earnings, after debt servicing, are too low to maintain the imports of spare parts and new equipment needed by its copper industry. Copper production is, therefore, reduced, which means that the shortfall on foreign exchange earnings increases. Since copper sales represent over 90 per cent of all Zambian export earnings the problems of the copper industry have a dramatic effect on the whole economy. These two examples illustrate the way LDC producers can slide into a vicious cycle of decline. Neither market mechanisms nor international political responses appear capable of arresting this cycle.

RENEWABLE AND ENVIRONMENTAL RESOURCES SCARCITY

Economic and cultural responses have also failed to arrest the depletion and degradation of a number of resources which have traditionally been regarded as plentiful and naturally renewable. These include forests, fish, plant species and the capacity of the environment to absorb and degrade waste products. It is increasingly clear that such resources are not naturally renewable in the full sense. Either rates of use have to remain at or below the natural regenerative capacity or this capacity must be artificially increased up to levels of use. If the balance between use and regeneration is not achieved, then the 'renewable' resources (including the quality of the environment) will be 'mined' as surely as any metal or energy stock. It is well known that over-use, pollution or habitat destruction has already resulted in the extinction of dozens of plant and animal species. The existence of hundreds more is threatened.

In a very real sense these problems of exhaustion or degradation are not physical in origin, but result from the failure of human beings to devise appropriate methods of management. They arise because some of these resources have traditionally been regarded as 'free'. As a result they have not been marketed (or by their nature cannot be marketed) as other resources have been. They have often

19

been regarded as common property resources. This term simply means that they have been conventionally freely available, for use by all, and are not owned and controlled exclusively for the benefit of one person, company or even nation. Clearly, if a resource is not marketed it has no market value or price. There is then no incentive to increase or improve its supply when scarcity problems arise. Likewise, if a resource is openly accessible to everyone, individuals are unlikely to take conservation measures, since any benefits will be reaped by all other users. No fisherman, for example, is going to reduce his own catch to conserve a species unless he can be sure everyone else will do the same. Still less is he going to spend money on fish hatcheries to increase the stock in rivers or the sea unless he can charge others for the resulting increase in stocks.

For these free, or common property, resources some form of governmental action is necessary to ensure that individual, short term, self-interest does not degrade or exhaust the supply. Even when the resource in question occurs within one country, the task of control is never easy. Let us consider the topical case in Britain and Europe of the degradation of water resources by nitrate pollution. For well over twenty years it has been known that excessive levels of nitrates in drinking water may be associated with stomach cancer and can cause the 'blue baby' syndrome. (This is a disease called methaemoglobinaemia, which reduces oxygen levels in the blood of babies.) It has also long been known that the two major sources of nitrates in water are nitrogen fertilisers applied to increase crop yields, and the ploughing up of grasslands (which releases nitrogen stored in the soil). The European Commission has recognised these nitrate risks. It incorporated a maximum acceptable concentration of 50 mg per litre in its Directive on Water Quality for Human Consumption. Britain accepted the directive, and with it the responsibility to ensure that all water supplies met the nitrate standard by July 1985. However, it is quite easy to pass a law, but much more difficult to enforce it (Figure 1.5).

The British government faces two options. Either it has to persuade, or force, farmers to curb fertiliser application and restrict the conversion of grasslands to arable farming, or it has to persuade, or compel, water suppliers to invest heavily in treating water before delivering it to the consumer. Neither option is problem-free. In the first place, there are administrative difficulties involved in attempting to control the activities of thousands of farmers scattered over the country. Second, all current farmers will be economically

Figure 1.5 Areas in breach of Community nitrate limits, England and Wales, 1987

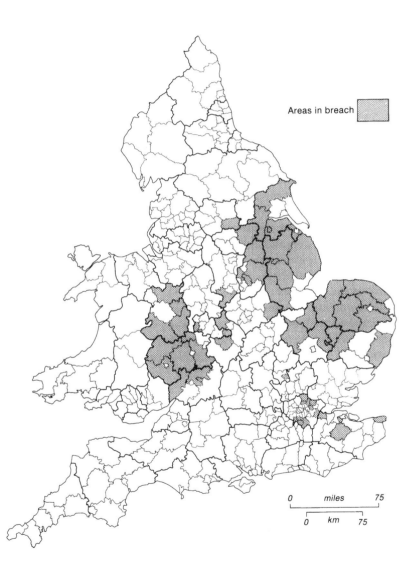

Areas in breach

disadvantaged, irrespective of whether they actually contributed to
the present nitrate build-up, since it can take over twenty years for
the nitrates to reach water aquifers. Such a time-lag also means that
reductions in fertiliser application or grassland conversion will not
necessarily solve present problems, although clearly they will ease
future difficulties. It is also possible that nitrate control measures
could lower yields and increase food prices. This could have the
undesirable consequence that poorer people would suffer more,
since they tend to spend a higher proportion of their income on
food.

These difficulties, plus the political power wielded by the farm
lobby (and its influential friend, the Ministry of Agriculture,
Fisheries and Food), have ensured that in Britain the polluter has
not had to pay for the damage caused. Rather the potential victim,
the water consumer, will have to bear the costs in increased water
charges for purification. While the water industry was publicly
owned, relatively little was done to reduce nitrate levels. In large
measure this was because the government was committed to reducing public spending and public sector borrowing. As a result, capital
expenditure on water services fell by 40 per cent from 1974 to 1987
and, hardly surprisingly, little investment in new treatment facilities
took place. Pressure from the European Community and from
public opinion has resulted in a shift of policy, and the required
investments will now be made; not now, however, by the public
sector but by the newly privatised water companies. These new
companies are not, of course, charitable institutions and they have
made it clear that improved treatment facilities will be provided
only if the costs can be recouped from increased water prices.

This example shows that measures to reduce pollution, or the
overuse, of renewable resources involve costs. Disagreement over
who should bear these costs is the greatest barrier to the prevention
of further degradation of renewable and environmental resources.
In the British nitrate case we have the problems of the distribution
of costs confined within one country. Even more intractable
difficulties will be encountered when a global common property
resource is under threat.

Perhaps the most critical scarcity issue currently facing human
society is the limited capacity of the global environmental system to
absorb 'greenhouse' gases. These gases are derived largely from
burning fossil fuels such as coal and oil. For at least thirty years
scientists have argued that we cannot go on assuming that the

environment will continue to act as a waste dump for carbon dioxide (CO_2) and other gaseous pollutants without adverse effects. Certainly, we cannot make this assumption when by destroying forest cover we are also reducing the capacity of the environmental system to absorb carbon dioxide. The tropical rain forests play a vital role in converting CO_2 into the oxygen on which all human life depends. Yet, every minute of the day, about eighteen hectares are either cut down for timber or burnt to clear land for agriculture.

The concentration of CO_2 in the atmosphere has risen from 270 parts per million (p.p.m.) before the industrial revolution to around 350 p.p.m. today. Methane concentrations have doubled since the beginning of the eighteenth century and carbon monoxide levels have doubled over the last hundred years. These physical facts are known. What is uncertain is whether these increasing pollutant concentrations are causing global warming. Neither do we know with any certainty what the effects of global warming would be. However, available data suggest that temperatures are rising (Figure 1.6) and that further warming could disrupt critical bio-geochemical cycles. Most governments now accept that risks exist, although their response so far has been stronger on rhetoric than on action. Some governments, including that of the UK, have preferred to delay decisions to control emissions until undisputed scientific proof of man-induced temperature increases has been gathered and until international agreements have been signed so that all countries take action together.

Controlling the discharge of CO_2 and other greenhouse gases will certainly be costly. Major improvements will have to be made in the fuel efficiency of vehicles; new technologies will be needed to reduce industrial energy use; homes and offices will have to be built to conserve heat, and rapid progress will need to be made in the development of non-fossil fuel energy sources. To stimulate such innovations it has been suggested that a new 'environmental tax' should be imposed on the use of fossil fuels and that strict pollution emission controls should be imposed on manufacturing industry, electricity generation and vehicles. However, such measures will inevitably (at least in the short run) increase manufacturing costs, increase inflation and reduce economic growth rates and standards of living. Why should any single country bear these costs when it cannot be sure that others will do the same? This is the nub of the problem.

Less developed countries are in an especially difficult position.

Figure 1.6 A century of global warming, 1880–1990

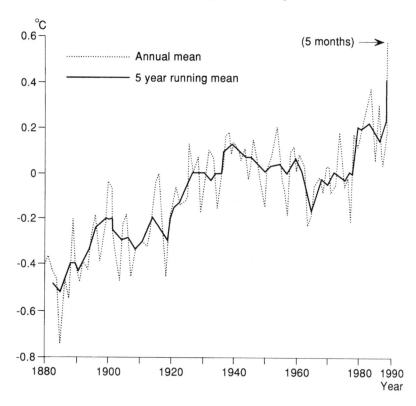

They would clearly suffer greatly if they were required to shoulder the heavy burden of control costs as they strive to raise their standard of living. Indeed, many countries, such as China, argue forcefully that the now industrialised and wealthy nations have created the problems by their prolific past use of fossil fuels, and the underdeveloped countries should not have their future growth hampered by costs which the advanced nations did not have to bear. Likewise, Brazil has argued that the conservation of its tropical rain forests would impose a major burden on its economy, while all other nations would gain the benefits. Understandably, many Third World states are prepared to make a contribution to solving the global warming problem only if the advanced nations are willing to give them help.

CONCLUSION

National self-interest, the preoccupation of governments with short term priorities, and the vested interests of so many individuals and groups in growth and capital accumulation, hardly portend well for the future sustainability of our global common property resources. The real problems of resource scarcity in the future are not likely to be shortages of materials. Market processes, as we have seen, contain powerful mechanisms to combat both the physical scarcity of 'developmental resources' and any geopolitical threats to resource supplies that may arise. But these same processes have not acted to protect the common property, the renewable and the environmental resources. Nor can they do so unless there is the political will to change management institutions and to alter the conventional values that have been placed on the environment.

QUESTIONS

1 How far will controls, such as those imposed by OPEC, be able to affect resource prices in the future?
2 Which kinds of resource scarcity are most easily solved by market forces?
3 What factors lead to changes in the value of resources over time?
4 Why do less developed countries find it so difficult to implement resource conservation?
5 Explain the difficulties faced in regulating the factors contributing to global warming.

FURTHER READING

O'Riordan, T. (1989) 'Contemporary environmentalism', in Gregory, D. and Walford, R. (eds.) *Horizons in Human Geography* (Macmillan: Basingstoke). Discusses the value issues in environmental concern, especially the problem of dealing with consumer goods.

Pearce, D. (1989) *Blueprint for a Green Economy* (Earthscan: London). A book that has become popular and influential in government thinking. Pearce has been an adviser on environmental policy to the government.

Rees, J. (1977) 'The economics of environmental management', *Geography*, 62 (4), pp. 311–24. A simple introduction to how the market works in respect of environmental issues.

Rees, J. (1989) 'Natural resources, economy and society', in Gregory, D. and Walford, R. (eds.) *Horizons in Human Geography* (Macmillan: Basingstoke). How society has to cope with resource and environmental questions – an up-to-date guide.

World Commission on Environment and Development (1987) *Our Common Future* (Brundtland Commission) (Oxford University Press: Oxford). A full review of the sustainability problems facing the world.

Zimmerman, E. W. (1951) *World Resources and Industries*, rev. edn. (Harper: New York). A useful reference text.

2

ENVIRONMENTAL HAZARDS

David K. C. Jones

The natural environment cannot be considered as neutral, merely serving as a backcloth for human activity. In reality there are complex two-way interactions between human societies and the natural environmental systems (atmosphere, hydrosphere, lithosphere and biosphere) which vary over time and space. As human societies have evolved, so their appraisals of the natural environment have changed, but such evaluations essentially focus on three main categories of interaction: environmental constraints, resources and hazards.

Environmental constraints (or biophysical limits) are met where the natural environment poses difficulties for human activity. Dense forests, swamps, steep mountains, aridity (in deserts) and harsh polar climates are examples of 'inhospitable' conditions. Such constraints can be overcome only through human organisation, technology and the heavy investment of labour and capital. There are innumerable examples of forest clearance, land drainage and slope modification which preceded contemporary activity. Such human interventions perhaps have their most dramatic modern manifestations in the irrigation schemes that have caused the deserts to bloom in parts of the United States, Libya and Israel.

In Chapter 1 *environmental resources* were defined as attributes of the natural environment that are valued by human societies at any point in time. But those *environmental events* that cause 'costs' to society by inflicting death, destruction, damage or disruption are usually referred to as *natural hazards* or *environmental hazards*. It is essential to recognise that both 'resources' and 'hazards' are human assessments or cultural appraisals and are, as such, intimately interrelated. Few phenomena can be deemed wholly good or wholly bad, the vast majority are combinations of the two. Thus 'resources'

27

and hazards should not be visualised as entirely separate groups of phenomena but rather as separate parts of a continuous spectrum. Whether phenomena are deemed beneficial 'resources' or damaging 'hazards' depends on the balance between their perceived 'costs' and their perceived 'benefits'. This, in turn, is largely dependent on the state of cultural development of the affected society.

In this chapter we will first examine six common misconceptions about hazards. The chapter then examines the global significance of natural hazards, the reasons for increasing hazard impacts and the various management options available to reduce the costs of hazards.

COMMON MISCONCEPTIONS ABOUT NATURAL HAZARDS

The overemphasis on disasters

Mention of the term 'natural hazards' generally invokes images of violent events – hurricanes, floods, tsunamis, earthquakes, volcanic eruptions, the collapse of mountain slopes – which result in widespread death, destruction and desolation. The media, and especially television, are increasingly dominated by harrowing images: the 1988 Armenian earthquake, hurricanes Gilbert and Hugo (in 1989), and the Californian earthquake (in 1989). Even the UK has had its share of destruction in recent years, with hurricane strength storms in October 1987 and January–February 1990, and major flooding by rivers (e.g. the Severn and at Maidenhead in 1990) and the sea (at Towyn, North Wales, in 1990).

It is undoubtedly true that very destructive events do take place, which sometimes have disastrous consequences for the societies affected. But the term 'disaster' is properly used only when the costs of an impact are very great. Disasters are, in reality, relatively infrequent (low frequency) large scale (high magnitude) hazard impacts. Such hazardous events are at one end of a continuum that includes numerous occurrences of medium sized impacts and large numbers of small events, many of which are so minor as to be inconspicuous and unnewsworthy. For any type of hazard, there-fore, it is important to examine its frequency of occurrence and compare it with the size of its impact. Such magnitude–frequency distributions show a few high magnitude–low frequency events (potential disasters) at one extreme, and a large number of low magnitude–high frequency events at the other, as shown in Figure 2.1.

Take, as an example, earthquakes. Each year 50,000–100,000 shocks are detected by sensitive instruments (seismometers). Of these shocks, perhaps 1,000 are large enough to be felt by humans, maybe 100 are of sufficient magnitude to cause costs, but only ten will be of sufficient size to be capable of causing catastrophic losses. Chance will dictate that only one or two may actually prove to be disastrous (see Figure 2.2). It follows, therefore, that emphasis on

Figure 2.1 Hypothetical relationship between the magnitude–frequency characteristics of hazard events and costs. There is a precise point at which a hazard event, by virtue of its size, becomes a disaster. 'Disaster' refers to the scale of the impact, or consequences, of a hazard event, not to its size, energy or violence

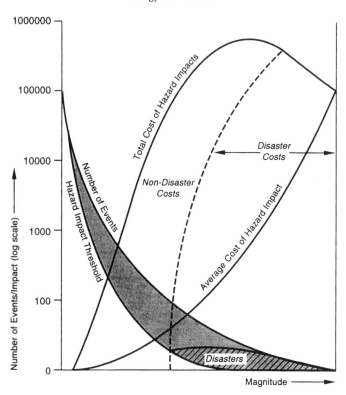

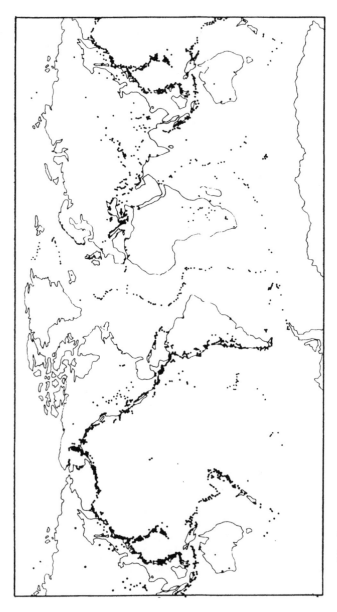

Figure 2.2 The location of significant earthquakes, 1970. Only one of these resulted in a major hazard (disaster): in Gediz, western Turkey, on 8 March 1970, killing 1,087 people. Several others caused minor impact (Figure 2.6). The distribution of earthquakes defines the two fracture zones which mark the edge of tectonic plates. The major concentration follows the continental fracture zone, which happens to coincide with many concentrations of people and industry

Source: Adapted from J. Whittow, *Disasters*, Penguin, Harmondsworth, 1980

the study of disasters may illustrate the most conspicuous consequences of hazard impact, but will inevitably lead to serious underestimation of both the distribution and the total costs of the hazard. For example, the British Isles have not figured in the global lists of earthquake disasters. Yet there have been sixty-one recorded damaging earthquakes since 1100 AD, including the 1884 Colchester earthquake which killed four people and damaged at least 1,200 buildings.

Human societies are powerless

The second traditional misconception is that environmental hazards are the consequence of the unpredictable behaviour of the natural environmental systems which cause impacts on innocent, unprepared societies. This view is embodied in the notion of natural hazards as 'Acts of God', and carries with it the idea of powerful events occurring randomly in time and space so that societies are defenceless. But all natural events have causes and patterns of behaviour which can be established through research. Rivers flood their valleys, coasts are inundated, volcanoes erupt, earthquakes recur and hurricanes return, so that patterns of recurrence can be established. The ever growing volume of scientific records makes it possible to identify the geographical limits within which particular hazardous events occur (the hazard zones) and to subdivide these areas according to the frequency of occurrence and/or the size of the events. From these data it is possible to establish the likelihood of events of a particular size occurring at a specific location over a particular time period.

The terms *recurrence interval* or *return period* are used to indicate the average interval between events of a particular magnitude. Thus the 100 year event, be it earthquake, flood discharge, windstorm gust, twenty-four hour rainfall, maximum or minimum temperature, drought, etc., is the value likely to be equalled or exceeded, on average, once in a hundred years. The probability of an event of a particular size occurring in any year is calculated by dividing the return period by 100; i.e. there is a 10 per cent probability of a ten-year event, 2 per cent of a fifty-year event, 1 per cent of a 100 year event, 0·5 per cent of a 200 year event, etc.

Viewed in this way, it can be seen that there are no freaks of nature, merely extremes. The infamous 16 October 1987 windstorm over southern England was not a freak, for there had been an even

Plate 2.2 Flood damage, Saudi Arabia, 1982. One of twelve bridges destroyed by a flash flood on a newly completed highway near Abha. The fact that such a large scale event occurred soon after the completion of the project is a perfectly normal attribute of 'return periods' (*Photo*: David Jones)

There are further magnitude–frequency characteristics of hazardous events that pose problems for human societies. The intervals between major impacts are frequently much greater than the time scales employed in political, planning and engineering decision-making. Thus it can be argued that very large scale hazard impacts are not so much unexpected as unanticipated. People tend to forget the previous disastrous events, preferring instead to focus attention on pleasanter times. (For example, spending money on a holiday or new car rather than on strengthening buildings or taking out insurance.) As a consequence, individuals, groups and societies become vulnerable to major impacts, and it is this vulnerability of society rather than the violence of nature that is frequently the fundamental cause of disasters. For example, both the 1988 Armenian earthquake and the 1989 Californian earthquakes were of exactly

33

the same size in terms of energy release (6·9 on the Richter scale) but 25,000 died in Armenia as compared with about sixty in the United States. These two events clearly show that the measure of physical impact reflects both the actual severity of the event and the state of human preparation for it. Put more bluntly, 'It takes acts of man to convert Acts of God into disasters' or 'Without people there can be no disasters'.

The naturalness of natural hazards

Primitive early human societies undoubtedly had little impact on their physical environment and were, therefore, affected only by hazards that were truly natural. Modern society is very different. While it is true that human activity has had minimal effect on the distribution and magnitude–frequency characteristics of most geophysical events such as earthquakes and volcanoes, the occur-rence of many other phenomena has been significantly affected by human societies. For example, deforestation has exposed the soil to erosion, thereby increasing gully development and landsliding; land use changes have dramatically altered the moisture budget and widely increased river flooding; winter sports increase avalanching; various atmospheric pollutants have combined to raise the acidity of rainfall; the burning of fossil fuels has increased global surface temperatures, influencing climatic events; and socio-economic pres-sures in many semi-arid areas have so overstressed the biosphere as to create desertification (see Chapter 3). Thus some phenomena commonly referred to as natural hazards are better termed 'human modified' or 'quasi-natural' hazards. As shown in Figure 2.3, some of these can be seen as human accentuated, whilst others are the sole result of human activity.

A good example of human-induced hazard is pollution. The dense, sulphurous smogs that used to blanket Britain's urban areas during the winter months chiefly resulted from domestic and industrial use of coal. The infamous London smog of 5–9 December 1952 caused 12,000 premature deaths. Fogs are a natural hazard, but the mixture of smoke and sulphur dioxide with natural fog results in a smog, which is clearly predominantly the consequence of human activity. Similarly, the Los Angeles photochemical smogs are wholly the product of chemical changes to motor exhaust gases in the atmosphere due to ultra-violet radiation.

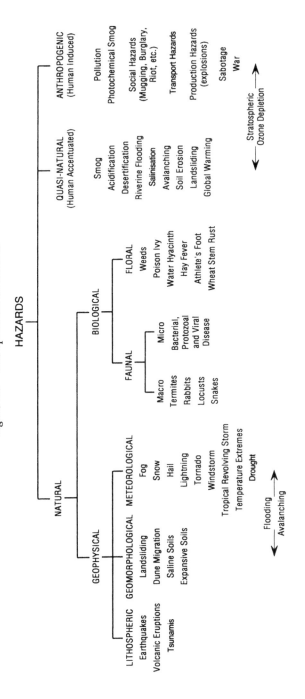

Figure 2.3 The spectrum of hazards

Overemphasis on violence

A fourth misconception is that natural and quasi-natural hazards are dominated by violent, high energy events. While this is undoubtedly true of many of the most conspicuous geophysical hazard impacts (e.g. hurricanes and earthquakes) it is not a universal truth. Drought, desertification, heatwaves, intense cold, fog, disease and biological pests all inflict costs in an insidious, non-violent fashion. These characteristics are well displayed by the global 'greenhouse effect' or 'heat trap' and the depletion of stratospheric ozone above polar latitudes, both of which are major current concerns about global environmental change.

Overemphasis on death

Death tolls have been overemphasised as a measure of hazard impact significance. It is easy to understand why casualty figures are widely employed in descriptions of disasters by the media: data on deaths and numbers hospitalised tend to be quickly made available, which is important in this era of rapid reporting; casualty figures are also more readily appreciated than other measures of impact (e.g. community stress indices); and historically the most obvious consequence of hazard impacts was undoubtedly the loss of life.

In reality there are fundamental problems inherent in the use of casualty figures in comparing the significance of hazard impacts. People do not have equal value, either within a country or when comparing states. A charismatic national leader is usually not thought to have the same worth as the humblest peasant, and a peasant in an area of labour shortage is not equal to an equivalent peasant in an area of overpopulation. Clearly age, life expectancy, skill and attitude are 'all important when attempting to measure the cost of 'life shortening' and this then has to be related to the total population. Take, as an example, the 1976 Tangshen earthquake, which killed an estimated 242,000 Chinese. This event had a relatively minor effect on the Chinese economy. The impact would have been greater had the death toll been in the nearby capital, Beijing, or in Hong Kong, San Francisco or Switzerland.

Death tolls are imperfect measures of hazard impact for further reasons. First, the timing of a hazard impact greatly affects the numbers of people at risk. An earthquake or hurricane at night with everyone asleep generally causes fewer deaths. Second, the variation in the state of preparedness of different societies greatly influences the resultant death tolls, as with the very different death tolls of the identically sized 1988 Armenian and 1989 Californian earthquakes. Third, some highly destructive hazard events claim relatively few lives; for example, the 16 October 1987 British windstorm killed only nineteen people but the insurance claims surpassed £1 billion sterling. Fourth, the risk of loss of life is not equal among hazards; some hazards, such as disease, may directly attack humans, while others may have only indirect effects on life expectancy (e.g. salinity, acid waters). Finally, it has come to be recognised in the developed world that casualty figures can be greatly reduced through the use of hazard forecasting, warning systems, evacuation and emergency procedures. These precautions have reduced deaths but have brought sharply into focus the escalating costs of destruction, damage, disruption and delay. Comparison of the two graphs in Figure 2.4 illustrates this point.

However, the evaluation of the costs of hazard impacts is not simple. It is possible to place financial values on many items such as the replacement of destroyed objects, the repair of damaged items, hospitalisation, burial, the creation of defences against the hazard and loss of production. But it is very difficult to place a value on the loss of life, the psychological effects, the stress, strain, anxiety, disruption of working practices, delay, or the loss of a cherished beauty spot or view.

The 'technocentric' fallacy

The final misconception concerns the ill-conceived view that technology can provide the means of total protection against hazardous impacts. This view is central to the 'society over nature' school of thought, or technocentrism.

Technology plays a complex role in the interrelationship between human societies and the natural environment. In many cases technology provides the means of defence against hazards in terms of both *structural* responses (dams, flood barriers, strengthened buildings, etc.) and *non-structural* responses (satellites to provide the basis for forecasts of severe weather, telecommunications to facilitate the issue of warnings, computers to work out insurance premiums). Conversely, technology can increase vulnerability to hazards in five main ways:

1. Technological development sometimes increases the likelihood of 'natural' hazard impacts by disturbing the natural environment.

Figure 2.4 Damage (left) and deaths (right) from hurricanes, United States, 1900–90. The totals fluctuate owing to the magnitude–frequency characteristics of hurricanes and the chance interaction between hurricanes and population centres. None the less the trends are clear.

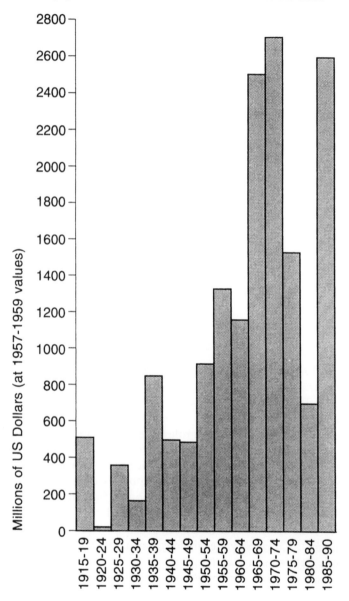

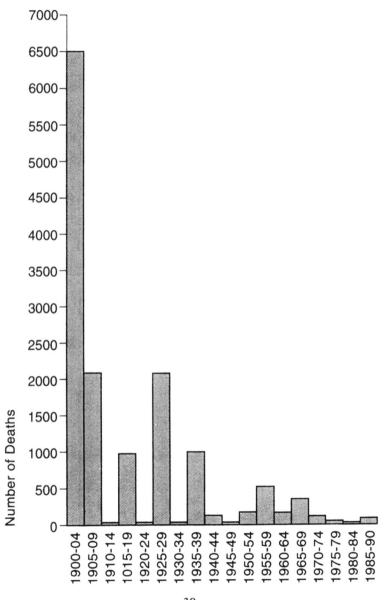

For example, thinning of the ozone shield due to the release of chlorofluorocarbons (CFCs) increases the potential for harmful impacts, including the development of skin cancers in humans.

2. The implementation of technologically based structural responses to existing hazards often results in populations gaining a false sense of security. This is known as the 'levée syndrome' and gains its name from the unfortunate situation that occurs when conspicuous defences such as flood walls, barriers or levées are constructed alongside flood-prone rivers. Once they have been constructed, it is generally assumed that all future river flows will be contained and that it is perfectly safe to occupy the adjacent low ground. In reality, all the barriers do is to raise the threshold at which flood overflow will occur. Short return period floods will be contained but longer period events (high magnitude–low frequency) will overtop the barriers and the consequences will be even more catastrophic. The same is true of 'earthquake-proof' buildings, where greater investment and greater durability only increase the threshold of damage but do not remove the risk altogether.

3. Technological changes frequently result in increased vulnerability to hazards. Fog probably had little effect on the stagecoach, but it plays havoc with motorway traffic and airline schedules; electric rail traction may be superior to steam locomotives in many ways but remains much more prone to disruption by snow, ice, freezing rain and wind; the introduction of modern steel and concrete buildings in certain coastal areas in the Middle East during the 1960s and 1970s resulted in unforeseen damage and decay due to attack on their foundations by aggressive salts.

4. Technology creates its own hazards. Most of the diverse human hazards are technology-based, ranging from conspicuous disasters such as the industrial explosions at Flixborough, Lincolnshire (1974), Bhopal, India (1984), and the Piper Alpha North Sea oil rig (1988), to the chilling statistic that 5,600 people were fatally injured on UK roads in 1988, bringing the total to near 300,000 in the last fifty years.

5. Technological development provides the potential for larger scale and more catastrophic impacts of hazards. This is displayed in its most extreme form by the forecasts that, following a nuclear war, there would be a 'nuclear winter'. Chernobyl, in 1986, provided an illustration of what could happen if a nuclear power station were severely damaged by an earthquake. In another case, the Vaiont Dam disaster in 1963 was produced by 250 million m^3 of rock

sliding into an impounded lake, causing waves of water up to 100 m high to overtop the dam, drowning 2,600 people in the valley below. Both examples show the dangers associated with the application of large scale technology.

IS THE WORLD BECOMING MORE HAZARDOUS?

There is little doubt that the impact of hazards is increasing rapidly. Data from various sources indicate increases in both the number of disasters per year and the annual cost of impacts (Figure 2.5). As we saw for hurricane damage in the USA (Figure 2.4), there is an upward trend in costs despite considerable expenditure on defensive measures. But perhaps the most remarkable data come from the analysis of earthquake impacts over the period 1900–79 (Figure 2.6), which reveals a year-on-year increase in the number of earthquakes that exceed a certian specified impact threshold.

There are three alternative explanations for the increase in frequency of earthquake impacts shown in Figure 2.6. First, it can be argued that the trend reflects progressively increased seismic activity within the continental fracture zone (see Figure 2.2). Such an explanation would need to be substantiated by evidence of increased number and/or size of earthquake shocks over time. Although there is some evidence of a minor increase, there is no scientific confirmation of a significant trend. Second, some of the increasing trend probably reflects better reporting and data collection since the 1960s. However, third, the increasing trend probably mainly reflects the growing vulnerability of human society due to population growth, urbanisation, industrialisation and infrastructure development. This indicates that increased losses from natural hazards are due, in the main, to inadequate adjustment by society rather than the increasing violence of nature.

The above statement applies to truly natural hazards. In the case of quasi-natural hazards it is also true that increased human modification of the environment has resulted in increased frequency and magnitude of some hazardous events (e.g. flooding) and has caused new hazards to emerge (e.g. desertification). That such activities will continue to pose problems in the future is clearly exemplified in the current debate on the potential impact of the greenhouse effect, especially the rise of 1–2 m in sea level which some have predicted. This would have a serious impact on many of the world's coastlines (e.g. Bangladesh), threaten cities such as

Figure 2.5 Losses from natural disasters, 1960–89

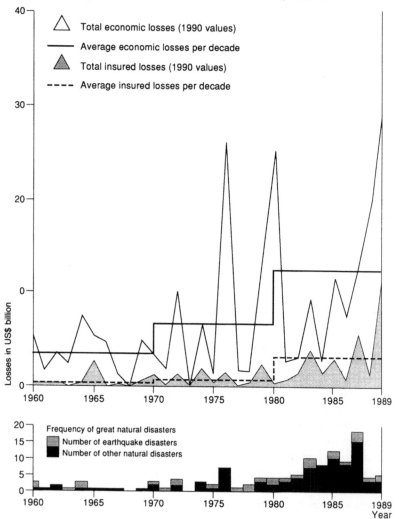

Source: Munich Reinsurance (1990)

Venice, London and Tokyo, and even make some island states uninhabitable (i.e. the Bahamas and the Maldives).

The 'greenhouse effect' or 'global heat trap' presents human society with a major dilemma for the future. There are plenty of gloomy predictions of catastrophic rises in sea level, expanding

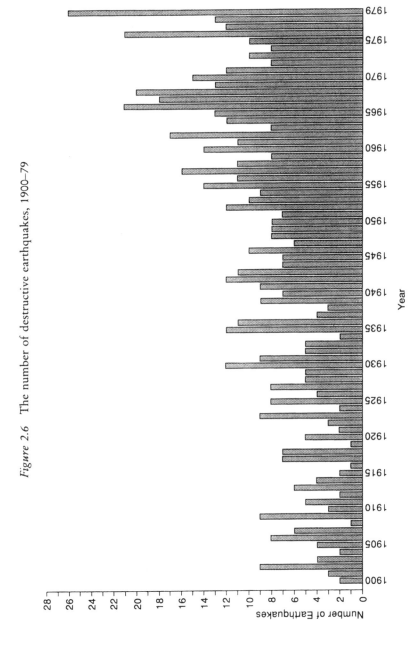

Figure 2.6 The number of destructive earthquakes, 1900–79

deserts, rapidly warming climates and increasingly violent atmospheric disturbances, which could create huge numbers of environmental refugees. But it is unlikely that the effects will be so major or so rapid. What is known is that atmospheric carbon dioxide absorbs terrestrial infra-red radiation, thereby reducing energy loss to space. The concentrations of atmospheric CO_2 (as measured at Mauna Loa, Hawaii) have increased from 312 p.p.m. to 350 p.p.m. over the period 1950–88 (Figure 2.7) and the concentration is currently increasing at 2–4 p.p.m. per year. As shown in Figure 2.7, these levels are greater than the range established for the last glacial–interglacial cycle, and it is known that the 1980s included the six warmest years on record globally. An extrapolation of the trend shown in Figure 2.7 produces hair-raising results. However, there are some fundamental uncertainties about these predictions, as discussed in Chapter 1. The need for scientific investigation of the mechanisms of greenhouse effects is clear. But the threat of sudden catastrophe is fantasy.

Overall, however, it must be recognised that increased losses due to hazards are predominantly the consequence of human activities. While some reduction in costs can be achieved by controlling or constraining natural forces, the greatest potential for hazard loss reduction lies in adjusting human activities so as to reduce vulnerability.

THE GLOBAL IMPACT OF NATURAL HAZARDS[1]

Major natural hazards have a global as well as national and local significance. For this reason the United Nations has declared the 1990s to be the 'International Decade for Natural Disaster Reduction' (IDNDR), stating that:

> During the past two decades, natural disasters have been responsible for about 3 million deaths and have adversely affected at least 800 million people through homelessness, disease, serious economic loss and other hardships, including immediate damages in the hundreds of billions of dollars.

These statistics are appalling and are probably an underestimate. Global annual death tolls of 150,000–250,000 are a generally accepted

---

[1] For the remainder of this chapter the term 'natural hazard' will be used to mean both natural and quasi-natural hazards.

Figure 2.7 Global atmospheric carbon dioxide concentrations. Data cover the last 160,000 years, based on antarctic ice core samples and recordings at Mauna Loa, Hawaii

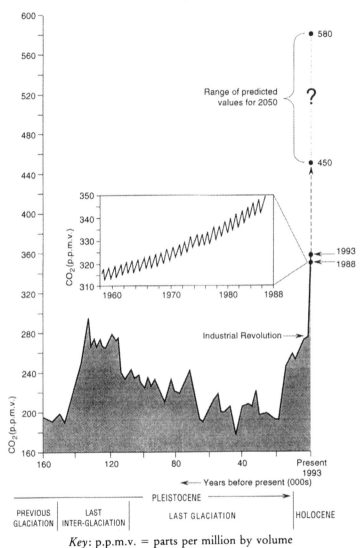

Key: p.p.m.v. = parts per million by volume

estimate. Of course, such average figures mask annual variation due to (1) the magnitude–frequency distribution of natural events and (2) the chance interation between natural events and human society vulnerability. As a result of these interactions, there are, every now and then, huge impacts – such as the Bangladesh cyclone floods of 1970 (225,000 killed), the 1976 Tangshen (China) earthquake (242,000 killed), the 1923 Kwanto (Japan) earthquake (143,000 killed, of whom over 60 per cent were burnt to death) and the 1556 Shensi (China) earthquake (830,000 killed).

While these figures are distressing and reflect major local disruption, how significant are they in global terms? The global population growth rate is estimated at 200,000 per day. About 20 million Russians were killed or murdered during the 1941–45 hostilities and a further 28 million under Stalin's agricultural collectivisation programmes. Even during a period of supposed global peace there were 133 wars which resulted in 25 million dead between 1945 and 1977. It is clear that 'man's inhumanity to man' is of greater significance in generating loss of life than are natural hazards, except in the case of rare, very great magnitude impacts.

Nevertheless, global natural hazards remain an important concern for two chief reasons. First, they impose a growing cost on society. Such costs are sometimes referred to as 'the natural tax'. The level of this 'natural tax' is uncertain. It has been estimated that the annual global cost in 1977 was US$40 billion, of which US$15 billion were expended on preventive measures (such as the Thames Flood Barrier). These figures undoubtedly underestimate the true scale of costs.

The costs of this 'natural tax' impose restrictions on the level of other economic activity because they direct resources to hazard limitation that could otherwise be used for other purposes. Populations continue to grow, while urbanisation, industrialisation and the complexity of economic activity all continue to increase. This means that the potential for destruction, damage and disruption is escalating and that the adverse effects of severe impacts can be spread ever further away from their primary site. These growing economic and social impacts of natural hazards will continue to increase the level of 'natural tax' in the future.

A second reason for concern is that natural hazards do not fall equally across the nations of the world, but are particularly severe in the developing world. This often leads to the conclusion that the countries of the developing world are more ravaged by natural hazards or hazard-prone. Inspection of maps (for example Figure 2.8) reveals that this is true only in part. While earthquakes, tropical

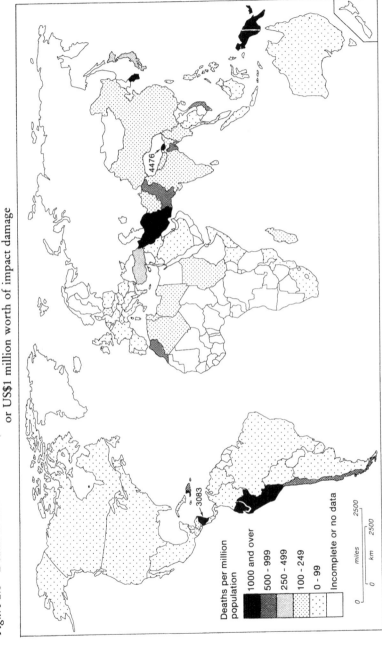

Figure 2.8 Deaths in natural disasters, 1947–73. Disasters are defined here as events which caused 100 deaths or 100 injuries or US$1 million worth of impact damage

revolving storms, volcanoes, insect pests and other hazards undoubtedly do heavily tax the Third World, they also severely affect countries like Japan, Italy and the USA. However, the number of deaths due to disasters is certainly highest in the most densely populated countries of the developing world. For example, the number of deaths per million local population for the period 1947–73 have their highest levels recorded in Bangladesh (4,233), Guatemala (3,378), Nicaragua (2,609) and Iran (1,521), although Japan (284) also scores surprisingly high.

Do these statistics indicate that the developing world is more hazardous? Only in part. Table 2.1 analyses the types of natural disasters. If drought and desertification are excluded, then 67–93 per cent of the deaths can be attributed to floods, tropical revolving storms and earthquakes. Thus the most dangerous parts of the globe are (1) those areas where steep, tectonically active mountain belts lie close to coastlines which experience frequent, intense storms (bringing earthquake, volcano, landslide, rainstorm and flood disasters), and (2) extensive tracts of low, flat land prone to inundation by major rivers and/or the sea.

Many developing countries have these characteristics, but so do

Table 2.1 Number of disasters by natural agent, 1947–80

Agent	Number of disasters
Floods	333
Typhoons, hurricanes, cyclones	210
Earthquakes	180
Tornadoes	119
Thunderstorms	37
Snowstorms	32
Heatwaves	25
Coldwaves	14
Volcanoes	18
Landslides	33
Rainstorms	33
Avalanches	12
Tidal waves	7
Fog	3
Frost	2
Sand and dust storms	3
Total	1,061

Source: Shah (1983) 'Is the environment becoming more hazardous?', Disasters, March, pp. 202–9

Japan and parts of the USA. It is difficult, therefore, to sustain the simple, deterministic, cause–effect argument that hazardous circumstances increase, or at least sustains, underdevelopment. What we see instead is that many developing countries are extremely vulnerable to hazards because they lack efficient forecasting organisations, warning systems, evacuation procedures and measures of physical protection. It is also true that significant populations within developing countries are especially prone to hazard impacts because they have been marginalised by development pressures. These are the poor, the dispossessed, the squatters, who live in extremely poor quality housing/shacks/shelters, often in dangerous locations such as steep slopes or floodplains. It is these people who are most vulnerable and who suffer the most.

Thus the 'natural tax' not only varies between countries, it also varies between groups within countries. It is this feature that has led some writers to refer to earthquakes as 'classquakes' because of their disproportionate effect on the poor. This can be summed up by the statement 'The rich lose money but the poor lose their lives.'

The relationship between deaths from disasters over the period 1947–73 and the income levels of countries can be seen from Figure 2.9 (where income is measured by GNP – see Chapter 4 for explanation). The countries can be neatly divided into four groups:

1 High deaths – high income: hazardous countries capable of sustaining losses because of great economic strength (Japan, Hong Kong).
2 Low deaths – high income: developed countries that are either not especially hazardous (Germany) or are well adjusted to hazard events in terms of 'life saving' but may suffer severe financial losses which they can sustain owing to their economic strength (USA).
3 Low deaths – low income: countries in the developing world which although poor are relatively free of hazard impacts, or underpopulated, or for which data are inadequate.
4 High deaths – low income: countries in the developing world where repeated hazardous impacts give a continuous need for relief and aid.

This fourfold division of countries reveals that the number, size, frequency and cumulative impact of natural hazards are not the only basis for assessing the natural tax. Hazard significance is also a

Figure 2.9 Deaths in disasters per million population, 1947–73, by national income (1973 US$). National positions will change with time. They will move upwards if nations suffer higher disaster death tolls, and to the left as they become wealthier

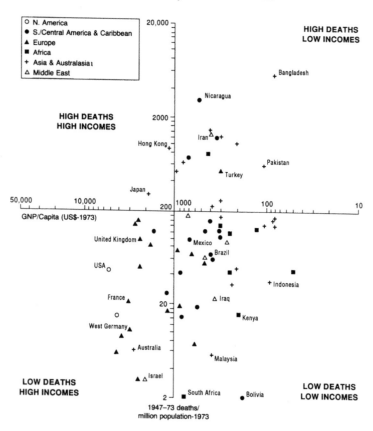

Source: Adapted from J. Dworkin 'Global trends in natural disasters, 1947–73', Natural Hazards Research Paper, no. 26 (1974), University of Toronto

function of the relationship between hazard costs and the ability of a country to sustain such losses. Most developed countries can sustain major hazard losses because of their economic strength or by spreading the costs through insurance (e.g. the 1989 Californian earthquake). By contrast, major hazard impacts have a crippling effect on developing countries, inhibiting the development process and necessitating additional injections of finance in the form of relief and aid. In many instances the impacts are so frequent and severe that hazard planning has to be seen as an integral part of the development planning process. Thus, although natural hazards cannot be seen to be the cause of underdevelopment they contribute significantly to maintaining the differential between the developed and developing worlds.

MANAGEMENT OF HAZARDS

There are many management strategies which allow the impact of hazards on society to be reduced. These responses vary in applicability, depending on the nature of the hazard and the technological status of the affected society. Figure 2.10 gives a classification of management action that distinguishes between measures directed to limiting the damage potential of the hazard ('active' or 'corrective' measures) and those directed to reducing the vulnerability of society ('passive' or 'preventive' measures).

For much of the time since 1945 emphasis has been placed on 'corrective' measures, particularly those directed to constraining

Figure 2.10 The range of available adjustments to natural hazards

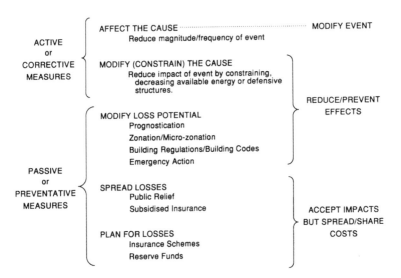

physical events. Technologically based modern society has naturally envisaged that technologically based engineering solutions (high-tech) could be used to effectively subjugate nature. Dams, levees, flood barriers, sea defences, wave spoilers, the artificial seeding of hurricanes, are just some examples. Most of these involve conspicuous expenditure on conspicuous structures, many of which turn out to be far more expensive and far less effective than was indicated in the original proposals (e.g. the Thames Barrier). It has become apparent that corrective measures undoubtedly have an important role in hazard loss reduction but are rarely an adequate solution on their own. They have to be used in combination with preventive measures which modify the vulnerability of the threatened societies.

Of the three groups of preventive measures, 'planning for losses' and 'spreading losses' are essentially concerned with reducing the financial impact of future events on affected individuals, groups and societies. They involve the acceptance that there will be future impacts but attempt to reduce the burden on individuals through some form of sharing. 'Planning for losses' involves the prior establishment of insurance facilities, the encouragement of individuals and organisations to take out insurance, and the establishment of reserve funds. 'Spreading the losses' concerns the ways in which the financial impact on those affected by adverse circumstances is mitigated through subsidised insurance and public relief. To be effective, both require knowledge of the magnitude, frequency and distribution of future impacts, especially for insurers, who set premiums based on their assessments of risk.

It should now be clear that all response strategies concerned with mitigating hazard losses need to be based on sound knowledge of the characteristics, patterns and periodicities of hazardous events. This scientific information has to be interpreted by those concerned with the human dimensions of the problem. This interface is best displayed in the group of preventive measures gathered together under the heading 'modification of loss potentials' (Figure 2.10). These represent the backbone of most hazard mitigation strategies and have at their core the assessment of future hazard through prediction.

Prediction is the assessment of the likely future distribution of hazardous events based on the study of their past occurrence. It assumes the uniformitarianist principle that 'the past is the guide to the future', an assumption that has to be modified with regard to quasi-natural hazards and those 'natural' phenomena that are influenced by medium and long term geophysical cycles. In essence it involves the recording of the location and magnitude of past events so that the extent of hazard zones can be established.

Study of the spatial distribution and magnitude–frequency characteristics of past hazards facilitates *zonation*. This is the division of areas into hazard zones and sub-zones which reflect varying levels of potential impact. This has become a very highly developed process in the USA, where, as shown in Figure 2.11, a process has occurred which results in *micro-zonation* of very small scale units.

Zonation and micro-zonation are essential to the effective implementation of many hazard loss mitigation measures. They provide the basis for deciding the scale of engineered defences, and form the basis for defining risk. Such work is therefore vital to the establishment of realistic insurance premiums. It produces the framework for land use planning in the sense that it provides the opportunity for organising patterns of land use so that vulnerability is minimised. This is known as *land use zonation* and essentially involves placing the most vulnerable land uses in the least dangerous locations and vice-versa. Thus, in a perfect world, hospitals, houses and nuclear power stations would be located in the least dangerous locations, while the most dangerous areas would be left as open space. Zonation also indicates the areas where special measures should be undertaken to strengthen buildings and other structures (building codes and building regulations), and shows how the measures employed should vary in extent, and therefore cost, over the area. Finally, it reveals the likely extent of future danger and thereby facilitates *contingency planning*, or the pre-planning of *emergency action* to be undertaken in the event of a disaster (emergency services, medical facilities, provision of temporary shelter, food, clothing, water supplies, alternative transport facilities, etc.). The effectiveness of an action is greatly increased if warnings can be issued based on monitoring of present conditions which are used as a basis for forecasting.

CONCLUSION

The preceding discussion has argued that the mounting cost of natural hazards is largely due to the inadequate adjustment of societies to the threats posed by the natural environment. This is a

Figure 2.11 Zonation and micro-zonation of earthquake hazard, United States and part of California. The broad national generalisations conceal major local variations. The micro-zonation map of California shows clearly those areas where precautions must be taken because of potentially intense ground shaking

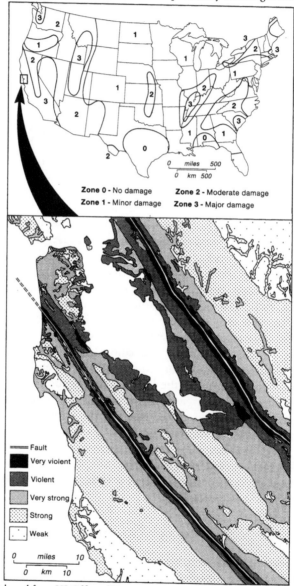

Zone 0 - No damage   Zone 2 - Moderate damage
Zone 1 - Minor damage Zone 3 - Major damage

Fault
Very violent
Violent
Very strong
Strong
Weak

Source: Adapted from US Office of Emergency Preparedness, and US Geological Survey

combination of human arrogance, based on the belief in the power of technology, and ignorance about the behaviour of environmental processes. It has resulted in many disastrous impacts and, therefore, any attempt to improve the situation should start with the 'human dimension'.

We have seen that there is a major distinction between the developed and developing worlds. Most countries in the developed world have political and planning systems that can respond to hazard impacts, scientific and technological skills capable of providing protection, and sufficient economic strength (including international reinsurance) to cope with the resultant costs. However, risk will continue to escalate because of economic growth, so it can be assumed that while losses of life will be progressively reduced to acceptably low levels, the scale of financial impact will continue to mount (see Figure 2.5). This is because hazard considerations (zoning, hazard-proofing) are frequently dominated by other, more powerful factors such as the historical legacy of land use, land values, land ownership, political considerations, locational advantages and profit maximisation.

In the case of the developing world the problems are rather different. In some instances, hazardousness is seen as a 'normal' condition that is too difficult to overcome. In other cases, hazard losses are just one of a number of major problems that impede development. Scientific and technical skills are often limited and adequate finance can usually be obtained only through costly loans or aid programmes, the latter often in the form of conspicuous engineering solutions that benefit the donor country. Land use controls are frequently rudimentary, insurance often virtually nonexistent and forecasting/warning services of limited effectiveness. Under these circumstances, death tolls and economic losses will inevitably continue to rise unless measures are taken internationally to reduce the impacts.

Here, then, is the goal of the UN International Decade for Natural Disaster Reduction:

to improve the capacity of each country to mitigate the effects of natural disasters expeditiously and effectively, paying special attention to assisting developing countries in the assessment of disaster damage potential and in the establishment of early warning systems and disaster-resistant structures where and when needed.

In other words, the UN has made a call for the developed countries to share their scientific knowledge and technical capability with the developing world, to help educate the threatened populations as to the nature of the hazards they face and the adjustment options available, and to assist in the provision of protective measures. One especially important element of the UN framework is the call for 'a shift in emphasis to pre-disaster planning and preparedness while sustaining and further improving past disaster relief and management capabilities'. In too many countries in the past, the emphasis has been placed on responding to disasters, usually with foreign aid. Such a 'back-to-front' and 'top down' approach is clearly unsatisfactory. What is needed is a 'bottom up' approach which starts with land use practices and builds up to include sophisticated technology where necessary. It should focus on avoidance and protection rather than aid repair, rehabilitation and reconstruction.

QUESTIONS

1 How do environmental disasters differ from environmental hazards?
2 What factors have led to the increase in hazards over time?
3 How do the costs of hazards relate to their frequency and magnitude?
4 What are the main factors leading to differences between countries in hazard impacts?
5 How can hazard prediction be used as a guide to policy actions?

FURTHER READING

Lane, F. W. (1986) *The Violent Earth* (Croom Helm: London). A readable review of the main types of natural hazards and their impacts on society. Represents an update of his earlier book *The Elements Rage* (1966), which is also worth reading.

Whittow, J. (1980) *Disasters* (Penguin Books: Harmondsworth). A wide-ranging consideration of the interaction between natural environmental systems and societies. Good examples of hazardous impacts.

Cuny, F. C. (1984) *Disasters and Development* (Oxford University Press: New York). A very good exposition of the problems of hazardous impacts on the context of development, focusing on the management problems.

Smith, K. (1992) *Environmental Hazards* (Routledge: London). A more advanced text.

3

DESERTIFICATION AND ITS MANAGEMENT

Helen Scoging

THE GLOBAL PROBLEM

Almost 700 million people live on the semi-arid and arid margins of the world's deserts, which embrace the poorest countries of the world. Desertification affects over 180 million people, and some 30 million km^2 of land, while over a third of the agricultural land in the world's dry zones faces desertification. In 1977 the United Nations Conference on Desertification recognised that desertification was one of the major environmental problems of the world and drew up a plan of action to identify the causes of desertification and to combat it.

Ten years later a United Nations Environment Programme (UNEP) report showed that desertification had increased and intensified, despite the efforts made since 1977. The area of land irretrievably lost through desertification continued to increase at over 6 million ha annually, while more than 20 million ha were annually being reduced to zero economic productivity. The rural population in areas severely affected by desertification increased from 57 million in 1977 to 135 million in 1987. Figure 3.1 shows that the areas of economic dryland affected by at least moderate desertification include 3,100 million ha of rangeland, 335 million ha of rain-fed cropland and 40 million ha of irrigated land. UNEP has termed this the 'time bomb of desertification'.

UNEP argues that a twenty year global programme to arrest further desertification would require an investment of about US$4·5 billion a year, with developing countries requiring US$2·4 billion annual expenditure. Despite the fact that aid continues to pour into these countries, especially following the increased public awareness fostered by newspaper and television coverage of drought and famine, the UNEP estimates of aid required remain well beyond the

Figure 3.1 The time bomb of desertification

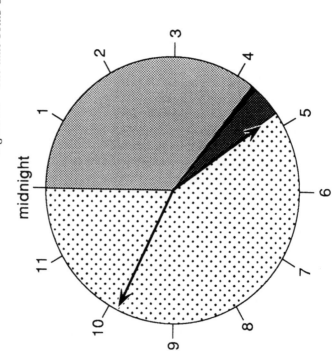

United Nations Environmental Program 1987

'the clock is ticking, an environmental timebomb. It is now 4.48 pm. At the rate of 27 million hectares lost a year to desertification or to zero economic productivity, in a little less than 200 years at the current rate of desertification there will not be a single fully productive hectare of land on earth.'

Land moderately or severely affected by desertification = 40% of the world's productive land

Rangeland (3100 million ha)

Irrigated land (40 million ha)

Rainfed cropland (335 million ha)

Undesertified potentially productive land (5215 million ha)

present levels of assistance. In 1980 the aid given was estimated by UNEP as US$600 million annually – only 25 per cent of the estimated requirement. The reasons for such inadequate support are many:

1 Most of the largest expenditures are needed in the poorest countries which are unable to tackle their problems without external financial and technical aid.
2 International assistance faces severe limitations because of political constraints in donor countries unrelated to the opportunities for cost-effective investments.
3 Economic returns on land rehabilitation are often spread over a long time and are low compared with alternative investments. Governments get more credit for projects with quick, conspicuous returns, and bankers using conventional discount methods have little interest in moderate returns thirty years away.
4 A major obstacle to greater investment is the difficulty of spending money productively. Desertification problems cannot be isolated from a complex web of local political, social and economic factors. Researchers can develop new seeds and rotation patterns for food crops, but applying these developments is very difficult when agricultural extension services are often non-existent. Government institutions are oriented towards export crops, and present marketing policies tend to depress grain prices for the benefit of urban residents.

THE WATER CRISIS

Water is a principal preoccupation for people on the desert fringes. The process of finding and obtaining water – digging into sand rivers, waiting for natural seepage to fill a gourd, and carrying it back to the village – may take up to sixteen hours a day. Very often, too, herds are taken to the water sources, so that livestock overgrazing, pollution and declining water levels may severely degrade areas of land surrounding them. In drought years, pressure on those sources that remain places an ever-increasing burden on the surrounding land. As a result, villages or water sources often mark the centres of expanding wastelands as vegetation is removed and soil is lost. As the pressure increases, the search for water and pasture for herds increases outwards from the village, resulting in a wave of degraded land spreading out from village centres. When these

Plate 3.1 Digging for water in Kenya (*Photo*: Helen Scoging)

degraded areas coalesce, as in a bad case of measles, much larger areas become exposed to irreversible changes in the environment.

THE FUELWOOD CRISIS

Like water, the increasing pressure of demand on fuelwood produces similar environmental problems. The main fuel in African countries, and in many other poor countries, is wood. Firewood provides less than 5 per cent of the energy used in Britain, but in developing countries as a whole it is more than 20 per cent, and reaches over 90 per cent in rural Africa. During drought periods, especially, the forests and woody species are reduced more rapidly than the regeneration period of tree species, while people have to walk further, covering a wider and wider area, for fuel. The chief alternative fuel in such areas is oil. But the price of this fuel rose very rapidly during the oil crisis years of the 1970s and early 1980s, thus increasing the demand for firewood. This made it profitable for lorry owners to cut large tracts of forest and transport fuelwood long distances for sale. The consequences of firewood scarcity and high prices are thus not only seen in the economic burden placed on the poor of a particular locality. As a result accelerating degradation of woodland has affected wide areas throughout Africa, Asia and South America, bringing large scale increases in the wind and water erosion of soil in deforested areas.

In India and much of Africa the reduction in fuelwood availability, and the rise in price, have resulted in the increased burning of animal dung for fuel. This dung was previously returned to the soil, providing a vital source of nutrients and organic matter to bind the soil. The nutrients wasted by dung burning in India alone equal more than a third of the country's use of chemical fertiliser. The consequent loss of vegetation results in losses of topsoil and damage to soil structure and fertility. Water infiltration into the soil decreases and run-off increases. This erodes the soils further, especially where they are fragile and subject to animal trampling and compaction.

THE NOMADIC PASTORALISTS

In the 1970s many observers blamed nomads and their herds for causing widespread degradation of land. Drastic herd reduction was seen as a possible answer. But the reality is more complex. Most nomadic groups have prudent systems of grazing control, designed to take advantage of available grasses without serious destruction. Over the last twenty-five to thirty years, however, the spread of settled agriculture has confined nomads into smaller areas with less reliable rainfall. This, together with the demarcation of national borders, has limited their freedom to move in response to changing pasture conditions. New boreholes have often disrupted traditional patterns of range-sharing among tribal groups, resulting in severe overgrazing around wells. Ironically, improved vetinerary services have also enabled herds to grow too large without the simultaneous improvement of grazing management.

Livestock numbers and movements do need better management,

but the flexibility and ecological knowledge of nomadism must be built upon rather than replaced. Officials, mindful of export earnings, ask how meat production and sales can be maximised without degrading pastures. But the nomadic family may have other concerns – how big does the herd need to be to ensure survival through the next drought? How many animals are required to meet the social and cultural obligations of marriage and other customs?

In Sudan's northern Kordofan Province, livestock numbers multiplied sixfold between 1957 and the mid-1970s. This put unbearable pressures on grasses and shrubs. As the population grew without a simultaneous transformation of agricultural technologies, the traditional cropping cycle – sound and sustainable when followed properly – broke down, resulting in declining crop yields and the outright loss of arable land. In the past, patches of land covered with *Acacia senegal*, a soil-renewing tree which also produces gum arabic and excellent fuelwood, were cleared and planted with millet, sorgum, maize, sesame and other crops. This occurred over a period of four to ten years. After that time the depleted land was left idle until the acacia scrub reinvaded. After eight years or so the trees could be tapped for gum arabic – a valuable cash crop – for six to ten years. Finally as the trees began to die they were burned and the cycle began anew.

This ecologically balanced cycle of gum gardens, fire, grain crops and fallow has now broken down. The 1968–73 drought was the final blow. Under the pressure of a growing population, the cultivation period has been extended by several years and the soil has been irreversibly damaged. Overgrazing in the fallow period prevented the establishment of seedlings. Gum trees were cut for fuelwood. Without a gum harvest for cash the farmers replanted their subsistence crops, thus further depleting the soil. As a result, in a wider and wider area, the acacia no longer returns after the fallow, but is replaced by non-gum-producing scrub.

THE CASE OF THE SAHEL

One of the most extreme problems of desertification has arisen in the Sahel (Figure 3.2). Since the 1970s, Mali and other countries in the Sudan–Sahel region bordering the Sahara Desert have been suffering from drought. In 1958 Mali had 450 mm of rainfall, in 1973 only 200 mm, and in 1984 only 100 mm. By 1985 Mali had experienced sixteen years of drought. The river Niger has been a

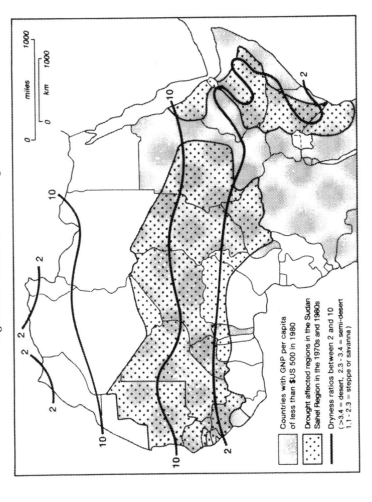

Figure 3.2 The Sudan–Sahel region

Countries with GNP per capita
of less than $US 500 in 1980

Drought affected regions in the Sudan
Sahel Region in the 1970s and 1980s

Dryness ratios between 2 and 10
(>3.4 = desert, 2.3 - 3.4 = semi-desert
1.1 - 2.3 = steppe or savanna)

great source of water and river transport for centuries, but by 1985 the river was at its lowest level since 1913. The country was short of 481,000 tonnes of food. There were virtually no harvests in northern Mali between 1980 and 1985, and herds were much reduced in numbers. By mid-1985 1·2 million people were on the verge of starvation. These prolonged periods of drought and food shortages are typical of the desertified desert margins.

Within Mali it is the north and east districts which are poorest and suffer most from food shortages. It is here that the Tuareg people, one of Mali's prominent tribal groups, are mainly concentrated. They are not settled agriculturalists but rely on pastoralism (nomadic livestock herding, particularly cattle and sheep, with some camels and goats). The animals provide milk and meat, as well as skins for shelter and clothing. Herders can exchange products such as butter and yoghurt for crops and other goods. Over the centuries pastoralism has adjusted to the fluctuation in the climate, with the herds migrating in search of new pastures. However, changes in this century have made it difficult for pastoralists to make the necessary adjustments. Population and herd numbers have increased, causing overgrazing and other pressures on land. These have added to the problems caused by drought. The 1970s and 1980s have thus threatened pastoralists with the destruction of their environment and livelihood. By 1985 the Tuareg tribe in Mali had lost as much as 70 per cent of their cattle. Forced to sell their animals to buy grain, some Tuareg people drifted to the towns, abandoning their hazardous way of life for the uncertainties of urban life. Others learned to cultivate crops for the first time and adjusted to sedentary agriculture, but very often on the margins of environmentally productive land.

After the poor harvest of 1984, emergency feeding centres grew around Mali's towns. By July 1985 the world had promised to give 253,000 tonnes of food aid. Of that 122,000 reached the areas where it was needed, the rest being stranded in West African ports, as there was insufficient money and resources for its transport. In Mali's Five Year Development Plan, 1981–86, over a third of spending was to combat the effects of drought and to raise food production, in addition to tree planting schemes to help supply timber for fuel and to combat desertification. The USA spent £150 million in Mali alone over the years 1974–84, but less than £600,000 went to schemes concerned with attacking the root causes of desertification. The European Community has been spending £60 million a year in Mali, but only recently has it begun a £300,000 soil conservation programme.

THE CAUSES OF DESERTIFICATION

The problems in Mali are part of a much wider problem which has affected over 30 million people in more than thirty countries in the African continent. The popular perception is that desertification is caused by the environmental pressure of drought combined with the human pressures of rapid population increase. This perception is not only a simplification of the complexity of the processes but is largely myth. Desertification should be seen, not as simply the advance of the desert, but as a process by which patches of formerly productive land become so degraded that their productivity is reduced to a minimum, and their natural regenerative capabilities are inhibited or destroyed. A cumulative process of decline then develops whereby people are squeezed out of these degraded lands on to new lands, which in turn become degraded. Such additional pressure, increasing the rate of degradation, may lead to the coalescence of desertified areas.

Myth 1: desertification is caused by drought

There are two conflicting views on the trends of drought. The first is held by those who believe that recent droughts are a manifestation of a long term climatic change: things are likely to worsen in the future. This theory is based on observations of the expansion of the cold air flow from polar zones. This is provoking a shift towards the equator of the major high-pressure belts which limit the advance of moist, rain-bearing winds from the equator to higher latitudes.

A second view suggests that, using evidence of water and lake levels going back at most 120 years, recent climatic fluctuations are no different from the past. This view holds that the data, at present, offer insufficient evidence to predict long term climatic changes. Rather, the evidence points to an irregular pattern of drought conditions in the Sahel over long periods of time: in 1910–15, 1944–48, 1968–73, 1983–85 and 1989. The process of desertification appears to be increasing because, imposed on the irregular drought pattern, are other factors which prevent regeneration and promote the positive feedbacks that increase the extent of environmental degradation. Thus human activity may create environmental conditions which not only maintain the potential for degradation (lack of plant regeneration) but also actually promote drought.

The effect of human activity on desertification derives, it is argued, from two main factors. First, at a global level, the burning of fossil fuels releases carbon dioxide into the atmosphere and increases the impact of the greenhouse effect in trapping long wave radiation from the earth. This contributes to increases in atmospheric temperatures – or global warming. This is a major issue confronting the world as a whole in its use of energy sources, but its effects are concentrated in the existing desert margins.

The second factor contributing to positive feedback is shown in Figure 3.3. The process is complex, but of great importance. The loss of vegetation through overgrazing and deforestation increases the surface albedo – the amount of incoming radiation which is directly reflected back into the atmosphere. Plants and litter conserve moisture, which has a low albedo, resulting in radiant energy being absorbed by the plants and converted into growth material, used in evaporation processes and the heating of surface soils. Convective precipitation is the major source of rainfall in drylands, with plumes of warm air rising freely until their moisture can be released from thunderstorm clouds. If the albedo is increased, less solar radiation can be absorbed during the day, and less heating will be available to induce convective currents. The result is not only reduced rain-producing processes in the lower atmosphere, but also reduced humidities and a lower incidence of clouds. This permits more infra-red radiation to be lost to space, thereby cooling the lower layers of the atmosphere. If such cooling processes occur over large areas, regional air mass subsidence occurs. This compresses the air, causing it to heat up and further reduce its humidity. The net result is a mechanism for suppressing convective rainfall over drylands. Perhaps more seriously, it is a self-perpetuating mechanism, since drought exacerbates overgrazing, which triggers desertification, which reduces vegetation, which increases albedo, which increases drought, and so on. This concept has been called 'biogeophysical feedback'.

Myth 2: desertification is caused by population growth

Population growth will certainly aggravate problems of environmental change. But it does not cause them. A more detailed

Figure 3.3 Positive feedbacks

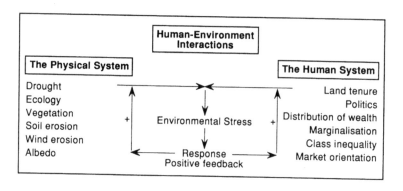

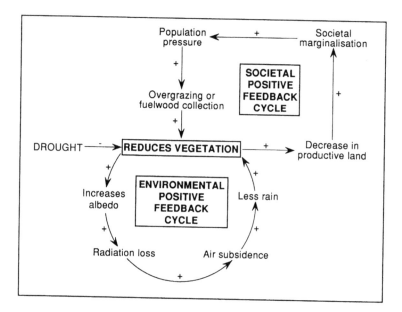

interpretation throws more weight on to the role of traditional agricultural practices, which have been too slow to adapt.

One feature of traditional societies, manifest throughout their history, has been their adaptability to change. Large families can be interpreted as a rational response to environmental conditions. Where death rates are so high that a quarter of all children die before their first birthday there is no motive for reducing family size. Children are an investment, providing parents with security against old age and sickness, and contributing at a very early age to the labour force. Currently there are no mechanisms whereby the advantages accruing to a nation as a whole through family limitation are reflected in the advantages to the individual family which chooses to limit the number of its children.

The more fragile an environment, the more delicate is the balance between the activities of people and ecological resources. However, the link between growth in number of people and desertification has too often been measured in terms of numbers without adequate qualification, leading some to argue that unchecked population growth will inevitably result in irreversible disaster. Malthus, in the nineteenth century, for example, argued that an increase in population to a level beyond the carrying capacity of the land must result in the elimination of the surplus population either by starvation or by other checks, such as warfare. The new version of Malthus's argument is that increases in population inevitably lead to the destruction of the land and that people, in order to avoid starvation, will move to other lands, which in turn become degraded.

An alternative view is provided by Ester Boserup. In her view intensive use of the land requires more people, not fewer. Rather than population growth being a check on economic growth, she argues that it is the maintenance of a traditional standard of living that is a fundamental impediment to agricultural development. This apparently academic debate on the direction of causal relationship between population growth and agricultural innovation is of major importance to the development of strategies to combat desertification. If Malthus's argument is correct, then every effort should be devoted to the restraint of population growth. On the other hand, if the lack of agrarian change reflects the lack of adaptation to population pressure, as Boserup suggests, then the prognosis is much more hopeful. Historically the evidence favours Boserup.

Myth 3: desertification is only a Third World problem

Desertification is a global problem, and although many of the countries suffering most severely are in the Third World, desertification is not exclusive to those countries.

At the simplest level the global dimensions of desertification include the movement of Saharan dust plumes across the Atlantic, or the migration of refugees across international boundaries. More profound are the international interactions between the developed and developing worlds in shaping the ability of the latter to enter world markets, limiting its potential for sustained economic development, and reducing its ability to cope effectively with the reduction of its resource base. This is exemplified in the impact of the rapidly increasing industrial and economic strength of developed countries on the developing world, from which many raw materials are derived. The developed nations consume such disproportionate amounts of raw materials, protein and fuels that it is impossible for the less developed countries to improve their standards of living. This has two major implications, as illustrated by the apparently endless demand for such commodities as meat and paper by the developed world. First, these demands have led to widespread clearance of the tropical rain forest, precipitating detrimental ecological, climatic and social changes and reducing vast tracts to permanent infertility. Second, the selling price of such food and raw materials is low compared with the selling price of manufactured products. So the developing countries suffer from an ever widening gap between their income and its purchasing power in world markets. They also bear the environmental brunt of a fast dwindling resource base and irreversible degradation.

Although media attention has most often been focused on the drought-ridden Third World – such as in sub-Saharan Africa, the Middle East, South West Asia, the Indian subcontinent and South America – the more developed and richer nations of the world have also been adversely affected by desertification. During the period 1978–83, for example, nearly 600,000 acres of fragile grassland in eastern Colorado in the USA were ploughed and converted to dryland crop production. By 1983, however, the high erosion risk of this land was already raising concern about a repeat of the Dust Bowl disaster of the 1930s. Again, across much of the United States the 1988 drought threatened significant areas with immediate soil erosion and loss of crops. A *Times* article of 21 June 1988 reported on a 'Bitter harvest of despair in the 1988 Dust Bowl', and compared conditions with those of the 1930s. In July 1988 the US government declared 850 Counties in eighteen States to be 'disaster areas' as a consequence of drought and poor farming practices. These are only two of many recent manifestations of desertification hazards in developed countries.

Thus neither the causes nor the processes of desertification are confined to the Third World, although it is there that society is least able to combat the problems. The developed countries are able to divert some of their resources into programmes to combat desertification, including weather modification, high-technology irrigation and drainage schemes, and economic subsidies. Poorer countries, and within them the extreme poor nomadic/pastoralist people, do not have such resources.

As environmental problems are primarily national in effect, if not entirely in cause, solutions must ultimately rest with the national decision-makers of the countries directly affected by desertification. As an example, the Sahelian states are confronted with a complex crisis made up of many elements. But this is the arena of development where some of the most subtle and intractable international issues raise their head. A country's ability to respond to desertification rests on its ability to balance its internal affairs with the pressures to enter world markets. Too often, however, when the rains come (and there is some reprieve from famine and the obvious symptoms of desertification) international – and in some cases national – interest in the sustained development of these countries disappears, and the problem is held over until the next drought.

We have seen that the processes of desertification are complex and involve a system of interactions between human activity and the environment, between the developed and developing worlds, and between the struggle of nations to enter world markets and their need to sustain agricultural production for their people.

MANAGEMENT OF DESERTIFICATION

Land management is the key to combating desertification. Management consists of using land in such a way as to repair degradation and to ensure that the capacity of the land is sustained for the future. We can distinguish between *avoidance* and *control* strategies. An example of an avoidance strategy is rotational grazing and shifting cultivation, which leave the reproduction of land capacity to natural repair forces and avoid intensive inputs on site. Control strategies, by contrast, seek to limit cropping or take land out of cultivation.

Land systems can be classified according to their sensitivity (the degree to which a given land system undergoes changes following human activity) and resilience (ability of land to reproduce its capacity after such activity). Resilience is a property which allows a system to absorb and utilise, or even benefit from, change. The character of land management depends on the two characteristics of sensitivity and resilience as shown in Table 3.1.

Sound development in poor rural societies suffering most from desertification generally requires new forms of social organisation. But advance is difficult because of the countless small, dispersed villages and hard-to-reach nomadic groups that are involved. This is why soil degradation is such an elusive problem for national governments and international agencies. It results from the actions of millions of individual farmers, each responding to particular economic and social incentives.

Table 3.1 Factors determining the character of land management

Resilience	Sensitivity	
	Low	*High*
High	Only suffers degradation under very poor management practices	Easily degrades, but responds well to management designed to aid reproduction of capacity
Low	Initially resistant to degradation but once thresholds passed it is difficult for management to restore capacity	Easily degrades, does not respond to land management, and should not be subject to human interference

In the Sahel, and many other regions, the quest for raw materials and foreign exchange has caused a near total emphasis on export crops such as cotton and peanuts. Cash crops can certainly play a constructive role in development. But, too often, the priority granted them by public policies has resulted in food shortages in times of drought, a vulnerability to worsening terms of trade, and ecological damage as the food crop area is squeezed. Perhaps the most serious consequence has been the associated neglect of research and infrastructure investment in support of agricultural systems for peasant farmers. This results in communities without essential public infrastructure, deprived of government services and short of capital for development and technology. But policies which might undermine the established economic primacy of the export sector would run counter to the large and powerful social groups which have a stake in the profitability of the export economy. So contradictions arise between management criteria for degraded lands and the dominant economic system.

LARGE DAMS

Large scale water supply and irrigation schemes have been one approach to attempting to overcome desertification. A number of developing countries are heavily dependent on large dams or barrages for irrigation and energy supply, notably Egypt, Sudan, India, Pakistan, Malaysia and China. In the Sudan, about 50 per cent of irrigation development expenditure in the early 1980s went on large dam schemes, a share rising to about 70 per cent in the Near East and the rest of Africa.

There is still debate as to whether one of the first large dams, the Aswan High Dam (1968), is on balance a success or failure, despite twenty years of experience. The building of such dams and impoundments is extremely costly and has frequently generated severe problems. Judged solely on technical and economic grounds, and not by the overall costs borne by the environment or people, most large dams have attained their planners' goals. For example, the overall contribution made by the Aswan High Dam to Egyptian agriculture is considerable and clearly outweighs the disadvantages. But the problems created by such large schemes are great in both environmental and social terms.

Environmental impacts may involve unforeseen effects. Water quality can be affected by the drowning of vegetated areas, resulting in an explosion of micro-organisms and leading to anaerobic or stagnant conditions. There may be large influxes of nutrient-rich waters draining fertilised catchments, resulting in algal blooms or aquatic weeds. Too often dams have been built only to find that the rate at which their impounded reservoirs silt up has been severely underestimated. For example, the Ksob Dam in Algeria silted up ten years before the irrigation project it supplied could pay for its construction. Water tables tend to fall downstream of a large dam, resulting in declining well levels, but tend to rise through reservoir leakage upstream, affecting structural foundations and soil salinity. The Aswan Dam has produced a 12 m change in water tables since it was filled in 1974.

Large dams also tie up considerable sums of investment capital and, too often, there are few local beneficiaries. Hydro-electricity tends to be transmitted to major urban areas or to foreign-owned industrial centres. Irrigation is more than likely to be for large scale commercial crop production with the profits channelled to the cities or to the multinationals. Flooding of reservoirs often involves the displacement of people, and alternative productive lands may be limited or unavailable. Although cash compensation is an alternative, it is difficult to distribute equitably and suffers from the risk that it will be misspent or poorly used. Lake Nasser, the reservoir impounded by the Aswan High Dam, covers 120,000 km² and cost over US$195 million (at 1974 exchange rates) for population resettlement in the Sudan and Egypt. The Volta project neccessitated the relocation of nearly 1 per cent of Ghana's 1968 population, of whom 67,000 were resettled and 9,000 were cash-compensated. The resettlement plans were ambitious, involving attempts to improve traditional agriculture and the creation of fifty-two new villages. By 1978, of the 67,000 relocatees transferred to resettlement villages only 26,000 remained.

The example of large dams illustrates the complexity of the management of desertification. Such dams have large scale impacts which are at the same time both desirable and detrimental. They are very often high on a government's list of management responses because they are so tangible, and because they can secure large loans and bilateral development aid from overseas. But such projects encourage dependence on external funding and expertise, and benefit the already more prosperous sectors of society in commercial agriculture, in industry and in urban areas. It is a very long time before these benefits filter back to the rural peasant farmer through the processes of structural development. Too often the displaced farmers have to re-establish life in resettlement areas which are often less productive and more marginal than those they left. Often the result has been a widening gap between rich and poor and increased desertification.

THE GREEN REVOLUTION

During the 1940s scientists from the Rockefeller and Ford Foundations in the USA were sent to help the Mexican government increase wheat and maize production through the use of improved seed, artificial fertiliser and other 'scientific farming' innovations. During the next decades a number of international agricultural research centres were established, such as the Consultative Group on International Agricultural Research, which began to disseminate the new high-yielding crop varieties (HYVs) of wheat, maize and rice. Results in Mexico seemed so promising that the term 'Green Revolution' was coined to describe the beneficial effect of HYV packages (seeds, fertilisers, pesticides and cultivation techniques) to

Third World countries. The term 'Green Revolution' implied peaceful, rapid improvement of agriculture through technological innovation and intensification of yields which would feed the hungry and raise farmers' incomes. Not only did it support the farmer, but the multinational agribusinesses which sold fertilisers, pesticides and HYV seeds also benefited.

The success or otherwise of the 'Green Revolution' remains debatable. On the success side, India, which in 1966 was the world's second largest importer of wheat, became in the late 1970s self-sufficient in that crop. But the Green Revolution bypassed large areas of the tropics, for example sub-Saharan Africa, where, in the 1970s, only about 15 per cent of the continent's agricultural production was cereals, and most of that was neither wheat nor rice. Unlike traditional crop varieties with considerable genetic diversity, HYVs lacked resilience, and occasionally and spectacularly succumbed to disease, drought or pests. The strongest criticism voiced against the Green Revolution has been that the packages were more easily or effectively adopted by the richer landowners and those with access to irrigation, with the effect that the gap between rich and poor was widened. Some of the problems associated with the Green Revolution arose out of unforeseeable developments. For example, during the 1970s, the sharp rise in petroleum prices caused by OPEC pushed the costs of inputs of fertilisers, fuel and pesticides beyond the reach of many farmers.

Some of the lessons have been learnt, and the Green Revolution continues, albeit in a different form. Recent innovations by international research centres have moved away from petroleum-based technology. Research is now more focused on improvements of indigenous species, along with local knowledge of plant species and livestock. New ways of controlling pests and weeds and improving soil fertility without recourse to expensive fertilisers and pesticides are also being developed. What is needed is that promising innovations be pieced together into a reliable and robust farming system which local people can accept. With this form of development the Green Revolution methods offer a great deal towards improving local output and land management in the desert margin.

SELF-HELP AND NON-GOVERNMENTAL ORGANISATIONS

Self-help schemes derive from attempts to decentralise strategies of soil and water management to families or communities. They are inevitably small scale and dispersed. A good example is in the Kitui district of Kenya. Here, earlier British colonial activity and, more recently, the influence of non-government organisations have improved water sources through large scale community rock catchments (which make use of the numerous bornhardts on the plains). One of the key problems remaining, however, is the inability of water abstractors to adjust to changing rainfall and water supply conditions, with the result that behaviour in periods of plentiful supplies is carried over to periods of drought.

The co-operation between non-governmental organisations in the area has also led to a scheme which places responsibility for water source protection in the hands of local people, drawing on their traditions of self-help (Figure 3.4). First established in local schools, and then through the employment of locally trained people to act as educators and skilled workers, the scheme seeks to supplement the community rock water supplies by the use of household roof catchments or rainwater harvesting. Simple and relatively cheap storage tanks and roof guttering can provide a household with adequate water within its own compound for its domestic needs for ten months of the year in seven years out of nine according to rainfall predictions. In four years out of nine a full year's domestic water supply can be harvested from roof catchments.

The scheme is particularly appropriate in a low density rural area where piped water supplies are unlikely, given the dispersed nature of the population and the very high costs involved. It illustrates the bottom-up approach to environmental management, where success relies entirely on local people 'owning' the problem and taking responsibility for its management. Aided initially by the provision of small amounts of non-governmental organisation funding for materials and training in building storage tanks and roof catchments, the scheme has developed quickly. Non-governmental organisations no longer do all the construction work, but pass it on to groups of local builders who, having been trained by the organisations, pass on their training to individual households and self-help groups.

This scheme has considerable advantages over the previous approach that had emphasised single, large scale watering sites. These led to large numbers of people and their herds concentrating around the community rock catchments which provide the only water supplies. This had a number of harmful environmental effects, including overgrazing by animals, wood cutting for fuel, trampling

Figure 3.4 Strategy for rural water supplies in Kenya

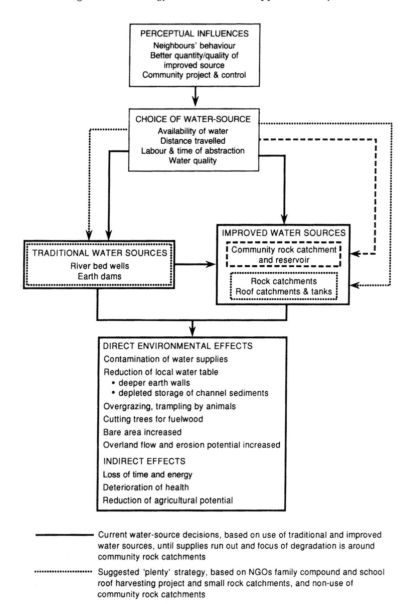

Plate 3.2 Desertification in Kenya. Overgrazed land round a water pipe which has failed because of falling water tables (*Photo*: Helen Scoging)

of the bare soil by animals, pollution of the water supply as animals and people abstracted water and washed in the source, and the attendant increase in potential erosion when the rains finally came. In addition there were the indirect environmental effects of the loss of time and energy spent collecting water under these conditions, time and energy which might otherwise have been spent in domestic or agricultural activity.

Rainwater harvesting by roof catchment is not without problems. In an area of highly unreliable rainfall, rainwater tanks are inevitably going to be empty for some parts of the year. This has two implications. First there is the need for a dependable community water source to meet the shortfall. Here the community rock catchments can be used, provided that they can be left unused when roof catchments are working. Second, there is a need to change attitudes so that an empty roof catchment tank is seen not as a failure, but as an inevitable and manageable event involving a shift in responsibility and source use. There is also the problem that the roof catchment scheme requires a level of education and hygiene which permits use of the additional water without incurring increased health problems. Waterborne disease, as a result of failure to filter roof water or to protect stagnant water from mosquitoes and hence malaria, would increase as more water becomes available unless adequate ancillary schemes were deployed. Finally, the success of the scheme may be discouraging national and regional government initiatives in improving water supplies.

CONCLUSION

Desertification is a complex problem arising as much from the social and economic characteristics of farming as from environmental pressures or reduced rainfall. Policies to limit or to reduce desertification are beset by difficulties. Much of the severe degradation of the Third World's marginal land is rooted in conditions and trends that the individual farmer can do little about. When economic development patterns and population growth force people to farm land that ought to be left in pasture or forest, the remedies lie outside the scope of on-farm technologies.

Some development schemes can themselves contribute to the problem of desertification. New wells and modern vetinerary services, coupled with a few good rainfall years, can increase herd numbers very quickly. At the beginning of a spell of dry years these large herds respond slowly to drought conditions. If the drought persists, large herds overwhelm the drought-stressed pastures, and one overgrazed area rapidly merges with another. Moreover, once the drought ends, animals can increase their numbers faster than rangelands can recover, thereby reducing the rate at which vegetation recovery can be effected.

Planting trees for fuelwood and building material will help combat desertification. Yet only 1·4 per cent of foreign aid to the Sahel in the 1980s went towards promoting such schemes. In some places reforestation will have to be increased eightfold by the year 2000 simply to meet estimated needs for household fuel. Niger is trying to introduce a national tree planting programme – some of which involves high technology (a plastic-piped, diesel-pumped drip irrigation eucalyptus fuelwood plantation scheme at Niamey) or the development of self-help tree nurseries, managed by groups of local people.

These strategies show the way in which improved understanding of the subtle links between the environment and society are leading to more sensitive management policies. Implementing such approaches remains among the main hopes, and challenges, for the future.

QUESTIONS

1 Discuss the interactions between society and the environment which promote desertification.
2 Why are desertification and its management global issues?
3 Compare the effectiveness of 'top down' and 'bottom up' strategies in the management of desertification.
4 How can we measure the effects of desertification?
5 Why is it difficult for the governments of countries affected by desertification to promote effective soil and water management?

FURTHER READING

Wijkman, A. and Timberlake, L. (1984) *Natural Disasters: Acts of God or Acts of Man?* (Earthscan: London). Covers a range of disasters, including droughts and desertification, with clear emphasis on the impact of disasters as social and political events which can often be prevented.

Eckholm, E. (1982) *Down to Earth* (Pluto Press: London). Wide-ranging discussion of soil erosion and management.

Grainger, A. (1982) *Desertification: How People Make Deserts, How People Can Stop and Why They Don't* (Earthscan: London). Excellent, frank, point-by-point arguments for social and economic causes and controls of desertification.

Heathcote, R. L. (1983) *The Arid Lands: Their Use and Abuse*, Themes in Resource Management (Longman: London). History and geography of human activity in drylands.

Whittemore, C. (1981) *Land for People – Land Tenure and the Very Poor* (Oxfam: Oxford). Clear, concise booklet on land and foreign aid in the developing world.

The New Internationalist. Monthly periodical covering wide range of issues concerning development and aid.

Sand Harvest (1987) computer programme (BBC Basic) and manuals (Longman Microsoftware), ISBN 0 58226211 9 (BBC disk). A superb teaching tool for role play and simulation of causes and effects and impact of human choices in desertification. Based on life in Mali, western Sahel. Three roles – Nomad, Government Officer, Villager – each with different preoccupations, but needing co-operation to save their environment and improve life.

4

GLOBAL ECONOMIC CHANGE

F. E. Ian Hamilton

FROM HOME-MADE TO WORLD-MADE

Economic life on our planet has changed very dramatically in recent decades. Thirty years ago, the world largely comprised quite distinctive national economies which were relatively self-sufficient in many goods. Their production systems were partly sheltered from each other by distance and the costs and time of transport. They were also divided from each other by the local ownership of firms and by national policies, such as trade barriers, to limit competition from goods made elsewhere. British students in the 1960s probably wore clothing made in Lancashire, Yorkshire or London, shoes from Northampton and hosiery from Leicester. They watched television on a set produced in Cambridge, Southend or London. At home were furnishings and electrical appliances produced in London, High Wycombe, South Wales or central Scotland. The parents ran a car which had rolled off an assembly line in Birmingham, Coventry, Dagenham, Linwood, Liverpool, Luton or Oxford. Boots of Nottingham supplied medication to cure many ailments. True, many of the raw materials embodied in these products were imported: cotton came from the USA or India, iron ore from Sweden or West Africa. But most intermediate manufactures required for them – like yarns, fabrics, steel, copper wire or transistors – were manufactured in British-owned factories, located in the United Kingdom.

The situation today is radically different. We live in a 'global village' in which momentous world events are transmitted instantaneously by television into our homes and products made abroad are integral to our daily lives. Shirts and blouses may come from Hong Kong or India, skirts and suits from Finland or Israel, pyjamas from Hungary or Portugal, anoraks from China, shoes from Korea or Italy. Students listen to hi-fi audio systems assembled in Korea, Singapore or Taiwan. Home furnishings originate from Scandinavia, Poland, Thailand or the Philippines; electrical appliances from Italy, West Germany, Singapore or Hong Kong. The family car is more likely to have been assembled in Paris, Turin, Antwerp, Wolfsburg, Munich or Valencia than in Britain. And far larger shares of intermediate goods are also produced abroad. Fabrics used by UK clothiers are imported from Hong Kong, Greece or Egypt. Japanese television sets are assembled in Britain from parts made in Asia. A Dagenham or Halewood-built Ford car contains parts which have been manufactured in almost all countries of Western Europe, the USA and Japan.

This simplified picture suggests some key global changes in manufacturing since 1960 and how they have affected what we buy. The so-called 'de-industrialisation' of the UK now seems well established. Not only did the number of jobs in industry virtually halve between 1970 and 1990, creating mass unemployment, but it also transformed Britain in the 1980s from a net exporter into a net importer of manufactures for the first time since 1700. The process also characterises some other old industrial regions (see Chapter 6) whereas many 'new' countries of the world, notably in East Asia, are attracting industry for the first time. Changes in industrial organisation are also seen in a shift from integrated 'national' manufacture to fragmented 'international' production. Such change means a replacement of coherent 'regional' industrial systems – like that of the former West Midlands motor industry – by output dispersed amongst plants in several countries. Closely linked with this trend, of course, has been a steep rise in the foreign or 'multinational' ownership of manufacturing and related service activity.

THE WATERSHED YEARS

Many long-established forces underlie these changes, but a series of events made the early 1970s a major watershed in the evolving world economy. First, microprocessors, which were invented in 1969, triggered an information technology boom, so stimulating the rise of new industries (such as the manufacture of personal computers) and services (such as computer software design) and the restructuring or decline of old ones like steel production. Second, the historic Bretton Woods agreement of 1944 that had established

81

the General Agreement on Tariffs and Trade (GATT) broke down in 1971, creating global economic uncertainty through the floating of currency exchange rates and oscillating international production costs. Third, a United Nations Industrial Development Organisation meeting in Lima (Peru) in 1974 called for the creation of a 'new international economic order' in which Third World industrial output would rise from below 10 per cent to more than 25 per cent of the world total by the year 2000. Some states sought to form international cartels of raw material producers as a way of raising export prices, increasing their revenue and funding development of the new order. The Organisation of Petroleum Exporting Countries, for example, pushed oil prices from US$3·00 per barrel in 1973 to US$30·00 in 1980. Last, but not least, pressure rose for resource conservation and environmental protection, further slowing growth in advanced economies. These and other major events during the 1970s and 1980s have greatly changed the global patterns of development and trade.

A DIVIDED WORLD

This chapter focuses on manufacturing industry, but a wider review of global change can first set the framework for this, and later, chapters. One measure often used to gauge a country's level of economic development is its gross national product (GNP). This figure represents the value of the output of all the extractive, manufacturing and service activities of a nation. The measure is approximate, but limits to its use need not detain us here. Figure 4.1, which maps GNP *per capita* for all countries of the world, shows the enormous gap between 'rich' and 'poor' today.

Much of Africa, Latin America, South Asia and the tropical island triangle of Indonesia, the Philippines and Papua New Guinea is desperately poverty-stricken. National income is below US$240 *per capita* (one tenth of the world average) in twenty-four states of sub-Saharan Africa and South Asia. Kampuchea (US$50 per head) and Chad (US$90) are poorest. Yet the wealthiest – the United Arab Emirates (UAE) (US$16,250 per head) and the United States (US$14,900) – lead a group of thirty-three states whose peoples enjoy *per capita* income levels which, at worst (the UK, with US$7,200, is in thirty-third place), are three times the world average. People in the UAE enjoy as much national income in a day as Kampucheans have in a year. Such great diversity suggests

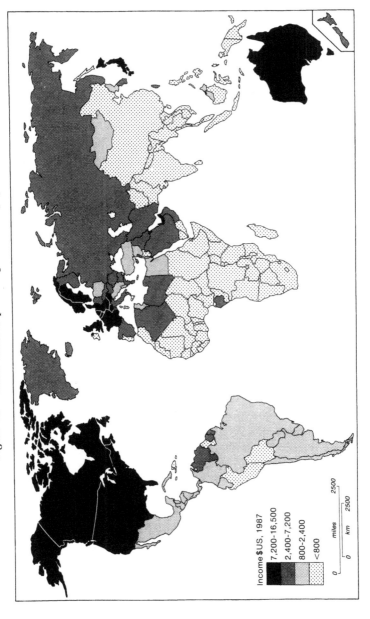

Figure 4.1 Gross national product per head of population

Income $US, 1987
7,200–16,500
2,400–7,200
800–2,400
<800

that many different economic 'worlds' coexist on this planet of ours.

One approach to generalisation has been the idea of 'three worlds' that emerged after the Second World War. A 'First World' comprised the industrialised free-market economies. Led by the USA, militarily its core embraced the North Atlantic Treaty Organisation (NATO), while economically it was linked in the Organisation for Economic Co-operation and Development (OECD). A 'Second World' of centrally managed, industrialising states under Soviet influence rivalled the First via the Warsaw Pact (WTO) group of countries and the Council for Mutual Economic Aid (CMEA or Comecon). A 'Third World' consisted of mostly excolonial lands which gained independence after 1945. Fragmented and vulnerable to border disputes, this Third World has been an arena for the First and Second Worlds to stage their economic, political and military rivalries. Such conditions gravely undermine efforts to shake off underdevelopment, especially in the face of rapid population growth and often adverse physical environments. Seven of the world's ten poorest states still suffer from conflict rooted in the First and Second World Wars – Kampuchea, Laos, Vietnam, Afghanistan, Chad, Ethiopia and Mozambique.

By the late 1970s this threefold division of the globe was felt by some observers to be inadequate. Economic trends were diverging considerably among Third World countries. A few of these states were becoming 'newly industrialising countries' (NICs), while others raised cash from natural resources, especially oil (OPEC), and became richer. These states were regarded as forming new Third and Fourth worlds while, elsewhere, rapid growth of population continued, and poverty, illiteracy and hunger worsened. This created a Fifth, low income, world which itself could be conceptually divided to leave a Sixth world of quite desperate poverty.

But this complex, if more realistic, view of the world has not been widely adopted, being overtaken by a broader, simpler, concept of two worlds: 'North' and 'South'. This was popularised by the Independent Commission on International Development Issues in 1980. The two worlds are well defined latitudinal zones, divided along an imaginary line which loops south to embrace Australia and New Zealand (Figure 4.2). Geographical location, however, was not the prime feature distinguishing the two zones, but rather their economic conditions. North and South are synonymous with 'rich' and 'poor', 'developed' and 'developing'. The former, including

Figure 4.2 'North' and 'South'

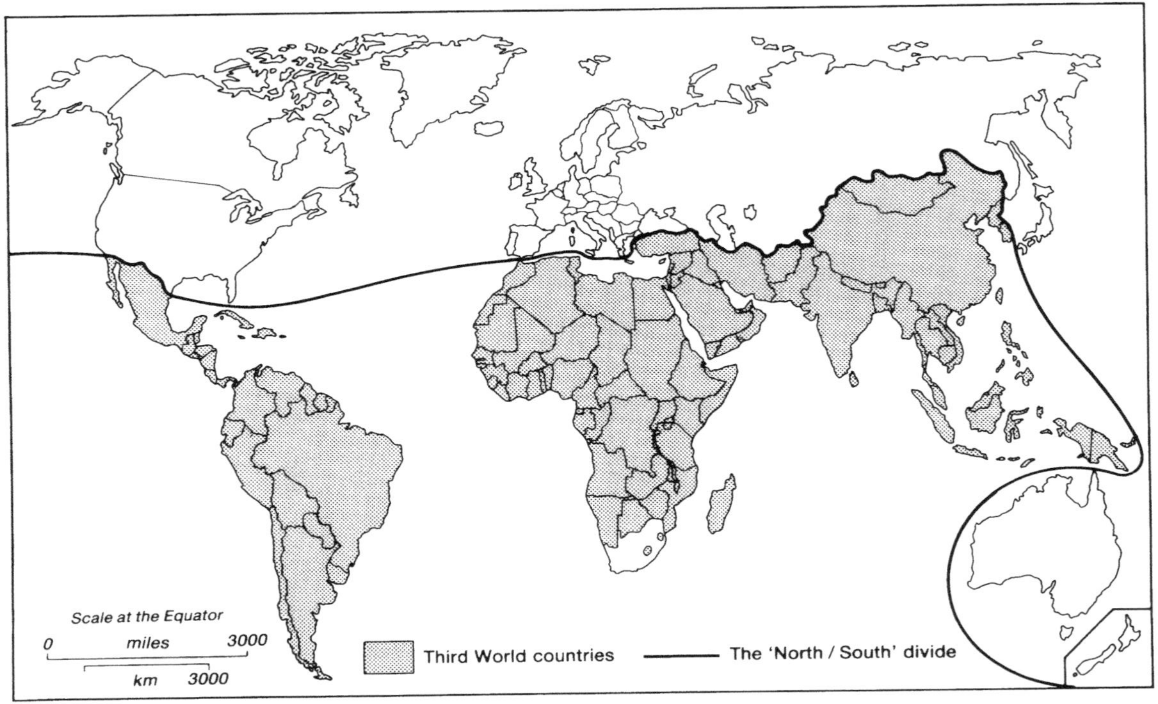

Scale at the Equator

0   miles   3000

km   3000

Third World countries

—— The 'North' / 'South' divide

Eastern Europe and the USSR, has a quarter of world population, four-fifths of its income and 90 per cent of its manufacturing. The latter contains three-quarters of the people but only a fifth of world income. Most new technology belongs to multinational firms headquartered in the North which

> conduct a large share of world investment and trade in raw materials and manufactures. Because of this ... northern countries dominate the international economic system – its rules ... regulations ... institutions of trade, money and finance.
>
> (Independent Commission on International Development Issues, 1980, pp. 31–2)

This focus on North–South served to make it clear that the East–West division, born of the Cold War, was diverting attention and resources from the solution of real global problems. Such a view is even stronger today when the Gorbachev era in the USSR has helped end, formally at least, the ideological, military and economic conflict between East and West. In essence, of course, discussion of how many worlds there are is not the key issue. What matters are the great economic disparities that lie behind the groupings. These disparities, in turn, depend to a significant degree on the varying role of manufacturing. Thus trends in national patterns of manufacturing industry are fundamental features of the changing world economy.

GLOBAL MANUFACTURING TODAY

Although economists, commencing with Adam Smith, have studied the nature and causes of national wealth creation for 200 years, policy interest in development came only after 1930 when the League of Nations advocated industrialisation in Eastern Europe to combat rural poverty. Subsequent work by the United Nations concluded that the expansion of manufacturing raised productivity and incomes by drawing labour from less productive farming, and fostering investments in services and community infrastructure. So 'industrialisation' became a major goal of governments in the 'South' right through three United Nations Decades of Development; 1960–70, 1970–80, 1980–90. Have the goals been achieved?

Table 4.1 ranks the largest sixty states by value added in manufacturing in 1985. ('Value added' is the difference between the value

Table 4.1 Value added in manufacturing in the sixty largest economies, 1985 (US$ billion)

Country	Value added	Country	Value added
1. United States	803·4	31. Bulgaria	9·9[1]
2. Japan	395·1	32. Denmark	9·7
3. Soviet Union	313·8[1]	33. Hungary	8·1
4. West Germany	201·6	34. Philippines	8·0
5. France	124·4	35. Norway	7·9
6. United Kingdom	101·5	36. Thailand	7·7
7. China	95·1	37. Saudi Arabia	7·6
8. Italy	93·9	38. Nigeria	7·4
9. Canada	58·9	39. Hong Kong	6·7
10. Brazil	58·1	40. Algeria	6·2
11. Spain	44·9	41. New Zealand	6·0
12. Mexico	43·6	42. Colombia	5·6
13. East Germany	40·1[1]	43. Greece	5·4
14. India	35·6	44. Pakistan	4·9
15. Czechoslovakia	32·3[1]	45. Singapore	4·3
16. Australia	30·7	46. Peru	3·4
17. Poland	28·4[1]	47. United Arab Emirates	2·7
18. South Korea	24·5	48. Ecuador	2·4
19. Netherlands	23·1	49. Morocco	2·0
20. Sweden	20·9	50. Kuwait	1·7
21. Belgium	18·6	51. Bangladesh	1·3
22. Yugoslavia	18·4[1]	52. Zimbabwe	1·3
23. Romania	18·3[1]	53. Libya	1·2
24. Austria	18·3	54. Tunisia	1·0
25. Argentina	18·0	55. Cameroon	1·0
26. Turkey	12·3	56. Ivory Coast	0·9
27. Finland	12·2	57. Bolivia	0·8
28. Indonesia	11·4	58. Sri Lanka	0·8
29. South Africa	11·1	59. Nicaragua	0·8
30. Venezuela	10·6	60. Dominican Republic	0·7
Total of sixty largest states			2,846·5
World total			3,300·0[2]

Note: [1] Estimated, but figures include mining and utilities.
[2] Includes the estimated value added in manufacturing in other socialist economies of Eastern Europe, Cuba, Mongolia, Vietnam and North Korea.
Source: World Bank, *World Development Report 1988*, New York: International Bank for Reconstruction and Development; Economist Publications, 1988 *The World in Figures*, London: The Economist, 1987

of the product as it leaves the plant and the cost of all parts, materials, etc., bought into the plant for the manufacturing process.) It shows the North's continuing dominance. Four states in the North – the USA, Japan, the USSR and West Germany – account for 52 per cent of world value added; the top twenty, 73 per cent. By contrast, the top four in the 'South', China, Brazil, Mexico and India, account for just 7 per cent, while the twenty largest southern economies provide only 11·6 per cent of the world total. Very little progress has been made, therefore, towards the 'new international economic order'.

Figure 4.3 presents these international comparisons of the value *per capita* that manufacturing adds. The pattern reflects differences in both the volume and the quality of manufacturing between states. Very few states – in North America, north-west Europe and Japan – exhibit value added exceeding US$2,500 per head. These are world leaders in output, productivity and innovation of high-technology products such as computers, electronics, chemicals and pharmaceuticals. Lower values of US$1,250 to US$2,500 per head imply developed but mature, more standardised industrial production in countries like the UK, Australia and New Zealand, the advanced centrally managed states like East Germany or the USSR, and NICs like Hong Kong. Most peripheral European states exhibit inter‑mediate manufacturing values (US$600–$1,250 per head), as do Korea, the Persian Gulf and Latin American oil producers. These nations tend to concentrate on lower value material processing (like steel, aluminium or copper) or labour-intensive functions (in clothing, shoes or simpler metal products). Meanwhile the low levels of *per capita* value added throughout the South reflects its limited industry, mainly the semi-processing of materials, or low quality finished goods fashioned by very poorly paid labour in much of Africa and South Asia.

In most industrial economies today manufacturing creates 25–40 per cent of GNP. This apparently low figure is explained by the high income generated in such countries by efficient agriculture and, especially, by high value services. Even so, the share of GNP contributed by manufacturing still remains a useful index of the level of economic development. It usually generates 10–20 per cent or less of GNP in underdeveloped Africa, the Middle East and Latin America. It accounts for more than 40 per cent of GNP only in nations which rely heavily for their income on a single resource for which they do primary processing, such as oil in Norway, Saudi Arabia or Mexico, copper in Zambia and Congo, or where, as in the centrally managed economies of the Soviet Union and Eastern Europe, governments have given priority to industry and defence over services.

Figure 4.3 Value added in manufacturing per head of population

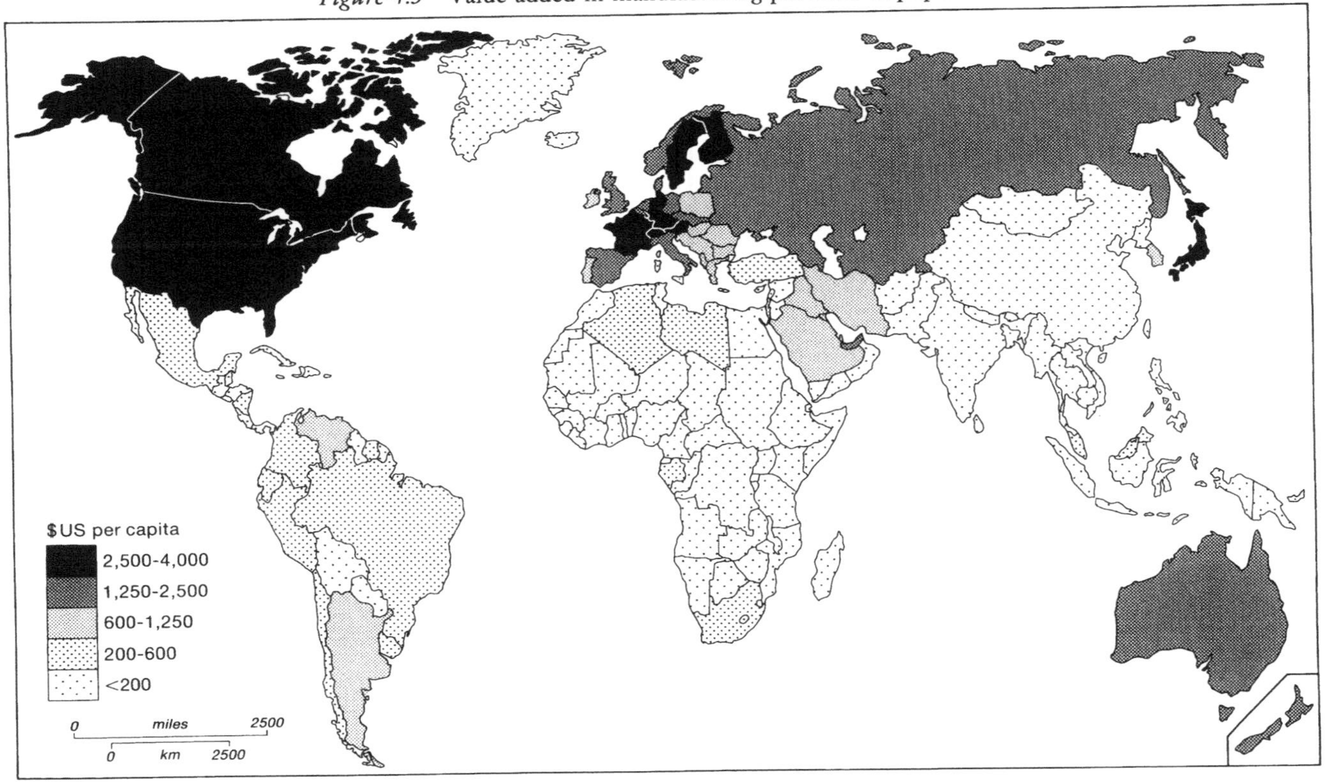

$US per capita

2,500–4,000
1,250–2,500
600–1,250
200–600
<200

0   miles   2500

0   km   2500

During the phase of global economic growth that ran from 1950 to 1975, sustained industrial expansion in advanced market economies was surpassed by even faster growth in centrally planned and developing countries. This began to narrow the global gap. But stagnation followed from 1975 to 1982, involving recession in the First and Third Worlds because of steep rises in the price of imported oil. Only centrally planned and oil-exporting states achieved growth as new factories built in the 1970s, using Western bank loans, came into production. Since 1983, however, a phase of *global divergence* has become evident and is typified by two marked features. First, 're-industrialisation' has been occurring in the North as the market economies, especially in Western Europe and Japan, recorded faster growth in manufacturing (1·5–6·0 per cent per annum) than either the centrally managed economies (2·0 per cent) or the Third World (0·8 per cent). Second, an eastward shift has been apparent, diluting the historic Atlantic-centred pattern and bolstering a Pacific-centred location of manufacturing. This is occurring because of (1) sharper differentiation along continental lines within the South as more Asian states – Thailand, Malaysia, China and India – join the 'Gang of Four' (Hong Kong, Korea, Singapore and Taiwan) in rapid industrial growth while output *per capita* actually declines in Africa and Latin America; and (2) a similar shift in the North as the Soviet Union and Eastern Europe stagnate, the EC grows weakly, and Japan and the western USA experience faster business expansion.

Many important forces have shaped these global trends: here, three – transport and communications, technology, and the growth of the multinational enterprise – are discussed briefly by way of illustration.

IMPROVEMENTS IN TRANSPORT AND COMMUNICATIONS

New types of transport underpin the growth, diversification and changing location of industry. Transport has become faster, cheaper, more flexible, able to carry more perishable and bulkier loads; and with the advent of lighter, compact and high value products such as pocket calculators or microcomputers, transport has become an 'enabling' rather than a 'determining' factor in location. Better communications, too, make cities, regions or nations more accessible to each other across the globe for all kinds of business transactions, improving contact between people for decision-making, and the exchange of information or money. This has major effects on industry.

First, such improvements have stimulated the geographical separation of distinct production stages and business functions from each other, and encouraged the growth of multinational firms to exploit appropriate locational advantages for each function. Metropolitan regions like New York city, London, Paris or Tokyo are specialising more in management and research; other regions in the North, the NICs and the Third World engage in factory processing. This 'new spatial division of labour' differentiates the former regions with national and global control functions, special skills and high incomes from the latter type with more routine output, work and lower incomes. No longer can geographical differences be explained by different manufacturing sectors alone.

Second, there are two particular transport innovations that have played a significant role in global manufacturing change. *Containerisation* has permitted quicker, cheaper intermodal transfer of freight between sea, road and rail – making industrialisation possible in places where traditional port docking facilities have been physically constrained. Hong Kong, for instance, has become the world's second busiest container port (after Rotterdam) by using barges to load and unload container ships out in the harbour waters. Second, the introduction of *long-haul jumbo aircraft* has made air transport a locational force, able to attract plants making light and valuable high-technology products, or firms offering producer services like computing, to sites on or near 'airport business parks'. This has fostered major new manufacturing growth in areas like the Thames valley west of London, near Washington, D.C., and in Singapore and Taipei.

Yet, third, faster communications now mean that time is worth more money. This is encouraging the use of 'just in time' delivery systems in manufacturing to reduce the need for, and costs of, holding stocks of materials or components. One result in both developed economies and NICs alike has been the clustering of suppliers – usually other manufacturers or warehousing firms – in central but less congested areas within a few hours' easy travel of major assembly plants. In Europe this sometimes involves international movement between factories sited strategically near national borders. Another effect is that satellite telecommunications are stimulating an expansion of global banking to profit from instantaneous data on curency exchange rates and company performance throughout the world. Before the advent of twenty-four-hour banking and share dealing, Hong Kong and Singapore were able to become second rank global banking centres (following London, New York and Tokyo) by trading in the hours when Tokyo was closed and before London opened the same day. Such growth of international financial services attracted foreign firms to invest in manufacturing in these and nearby Asian NICs like Taiwan, Indonesia, Malaysia and Thailand.

TECHNOLOGICAL CHANGE

Technology affects manufacturing in many ways. Directly, it yields new industries making products like computers or disc brakes, and new processes like oxygen steelmaking which can change existing industries and their location. New products can themselves become processes: motor vehicles revolutionised transport, and computers are doing the same to society. Indirectly, innovations alter the materials, capital or labour used in production, the character of manufacturing firms, and the locational attributes required for their survival or development. Two concepts related to technology have commanded attention recently: *Kondratiev waves* and the *product life cycle*. Both assist in understanding global change.

In 1925 a Russian, Nikolai Kondratiev, claimed that 'long waves' of fifty to fifty-five years' duration typified the growth of market economies. In the 1930s an Austrian, Joseph Schumpeter, showed that such waves reflected 'innovation cycles'. New technologies 'energise' each wave, yield groups of new industries, and may shift production advantage to new places. Thus the first Kondratiev wave (1770–1825) followed the innovation of techniques in Britain to use coal for heat to make better iron, and to make steam to power textile spinning and weaving machines. It moved manufacturing from scattered workshops to factories sited on coalfields. Application of steam to rail and sea transport then ushered in a second wave (1825–80) in Britain, creating new engineering industries both on and off coalfields in places like Crewe, Derby or Swindon. A third wave (1880–1930) came mainly in the USA, Germany and France with the invention of electricity, the telegraph, the telephone, the internal combustion engine, and the development of oil-based chemical manufacture and photography. Leadership slipped from Britain to continental Europe and North America. A fourth wave (1930–80) emerged in the USA, Germany and the UK with audio-visual, aerospace, synthetic fibre, electronic and petrochemical industries.

We now live in a fifth 'innovation wave' which is creating a 'thoughtware' or 'information' economy which articulates brain power by research and development (R&D) in such fields as producer services (like computer software), biotechnology and robotics, and involves close links between costly R&D and manufacturing. Japan rivals the USA for leadership in these new activities. Recession after 1974 resulted from a coincidence of energy price rises and the transition from the fourth wave to the fifth. The closure in developed countries of many factories was in part due to competition from new producers in the NICs and Third World, but it was paralleled by a continued growth of services. People argue that all this heralds a 'post-industrial society'.

The Kondratiev cycle theory attempts to contribute to the understanding of the development of entire economies. The product cycle theory, by contrast, is concerned with the fortunes of individual industries. Product life cycle theory postulates that a new product will be innovated and developed in an advanced economy, like the USA or Japan, where the necessary skills, capital and support services are available. As world demand grows for the product, its output will require standardised production techniques, larger factory space and cheaper semi-skilled labour so as to lower costs. This enables a diffusion of production elsewhere, at first, perhaps, to places like Western Europe where know-how and markets exist, later to NICs or the Third World with their low cost labour. Such 'maturity' of the product helps explain the global spread of the industries of the first, second, third and fourth 'long waves' (textiles, shoes, metal, plastic or electrical goods) to the latter types of locations. But it also explains why new fifth wave activities are causing re-industrialisation in the North.

MULTINATIONAL ENTERPRISES (MNEs)

World industry comprises a 'population' of private firms such as Ford or Toyota, state-owned enterprises like Renault or the Soviet Ministry of the Automobile Industry, and co-operatives. All private firms start small, often as family businesses. Many die, but some survive and expand to medium size while a few become very large corporations with many shareholders. The tendency to concentrate business into larger and larger firms is very strong in market economies. In 1909 the top 100 manufacturing firms in the USA and UK produced 22 and 16 per cent of national industrial output respectively; by 1970, 33 and 41 per cent. Schumacher's contradictory idea that 'small is beautiful', however, spread after 1973 and coincided with two developments.

The first linked the size of plants and firms with technological and market change. The transition from the fourth to the fifth Kondratiev wave involved plant closure and job loss in old, large scale industries and the opening of small firms in new activities. This appeared to be most marked in the USA, where the share of national output from the 100 biggest firms dropped to 24 per cent in 1980. In the UK it slid only slightly, to 39 per cent, thus expressing slow adjustment there to the fifth wave.

The second development was connected with public policy. Much established regional development policy in the Third World and Europe – which had been based on large, highly capitalised plants like steelworks or petrochemical complexes – failed to sustain economic progress in lagging or problem regions, often leaving them with so-called 'cathedrals in the desert'. This precipitated policy shifts to tap the 'indigenous potential' of small firms, i.e. to enhance the role of local enterprise in so-called 'self-reliant' development in the problem regions.

Yet, despite the recent boom in small enterprise, the proliferation of MNEs has accelerated throughout the First and Third Worlds during the 1980s. Firms become MNEs when they produce in two or more countries. This requires capital export, which is usually termed 'foreign direct investment' (f.d.i.). MNEs have their origins long ago in the trading companies that spearheaded European colonial expansion after 1600. But manufacturing MNEs emerged with the third Kondratiev wave: with large electrical or chemical firms after 1880, motor vehicles or camera firms after 1920. The first really big surge of MNE expansion, though, followed the US Congress's removal of restrictions on capital exports from the USA in 1955. American foreign direct investment flowed into Europe and Latin America in steadily increasing volumes. World f.d.i. rose from US$14·3 billion in 1914 to US$63 billion in 1960, US$386 billion in 1978 and US$950 billion in 1988. Giant MNEs have thus grown to be an integral part of a current trend towards the 'globalisation' of production organisation. They currently engage some 65 million workers worldwide, 3 per cent of world labour.

Why produce abroad? Some factors operate at home to 'push'

firms to locate plant overseas, while other factors 'pull' them to host countries abroad. Market trends (demand) and cost factors (supply) operate in both cases. The General Agreement on Tariffs and Trade, and 'customs unions' like the EC, freed global trade in manufactures, so intensifying competition: the phenomenon of import penetration of home markets became far more common both from within the North and from NICs. This 'pushed' firms to 'retaliate' by inter-penetrating each others' markets and finding sheltered markets in the South. Thus British firms moved output abroad as sales by foreign manufacturers rose sharply in their domestic UK market, while Japanese, West German, Swiss and American firms also built or acquired plants in Britain.

From 1979 to 1986 the forty biggest UK firms raised the proportions of their workers located overseas from 34 to 44 per cent – making some 415,000 workers redundant at home and creating 125,000 new jobs abroad. Yet, besides import penetration, firms perceived a deterioration in UK demand and supply conditions. High and rising unemployment squeezed home markets. Labour costs rose, strikes disrupted production, and unions resisted the introduction of new technology, work practices and sharper competitiveness. High interest rates made investment and stock-holding costly. All contributed to Britain de-industrialising and 'pushed' firms to invest abroad in multinational ventures. In contrast, rapidly appreciating values of currencies mainly explain shifts of output abroad in the early 1980s by firms like Brown-Boveri or Bally from Switzerland, Siemens or Bayer from West Germany, and Sony or Toshiba from Japan.

Plenty of 'pull' factors also exist to encourage firms to locate production in host countries abroad. Untapped and growing markets have attracted American, European and Japanese firms to set up production in both North and South. More, and rising, proportions of f.d.i. flowed within the North after 1960, as it offered big, vigorous markets and benefits from freer trade within customs unions like the EEC, EFTA, US–Canadian and Australian–New Zealand trade pacts. Thus American MNEs like General Motors or Ford, component suppliers Borg-Warner or Goodyear and oil majors like Texaco, Mobil or Exxon brought new manufacturing capacity to Europe in the 1960s. In the 1980s their Japanese rivals, such as Nissan or Sumitomo tyres, followed suit. On the other hand, Volkswagen, Honda and Toyota built in the USA. Recently, the high profits achievable from fifth wave functions and the availability of capital and skills have induced US information technology firms like IBM, Hewlett-Packard, Digital Equipment Corporation and Data General to establish plants and R&D facilities in the EC, while Japanese firms, like Fujitsu computers, have opened facilities in the USA. The UK, also, more recently began to gain f.d.i. as business confidence returned and at least some foreign-owned MNEs perceived it to be a low wage offshore platform from which to supply continental Europe.

Before the oil and debt crises of the early 1970s, Third World governments fostered inflows of foreign manufacturing to serve national markets behind protective tariff walls. This brought car assembly plants by Peugeot, Fiat, VW, GM and Ford to Latin America, for example. After 1970, cheaper energy and labour became the prime motive for investing in the South, a trend augmented by the introduction of free enterprise zone concessions and policies for the export orientation of output. MNEs responded by building plants as 'export platforms' to supply finished goods like Sony radio-cassette players from Singapore or Fairchild semi-conductors from Hong Kong to markets in the North. Capital did not flow everywhere: it avoided areas of high political or military risk in Africa, the Middle East, Indochina or Central America, and concentrated in stable hosts like the Gang of Four, Zimbabwe or Argentina. The establishment of cartels or resource nationalisation in the Third World often drove MNEs out of their traditional resource extraction activities and this, along with restructuring into fifth wave activities (like Exxon into computers), led to the stagnation, even decline, of MNE investment in parts of the 'old' Third World – like West Africa, the Middle East or Indonesia.

So the geography of MNE activity is very dynamic. Since about 1975, Western Europe has displaced the USA as the main global source of outward f.d.i. Japan ranks third, but is rapidly threatening to surpass both the USA and Western Europe in this respect. Western Europe and the Third World have both recorded fast growth in inward f.d.i., but Japan still hosts very little. Yet perhaps the most spectacular growth in MNE activity is in the USA, especially from Japan, which registered the fastest rise in outward f.d.i. after 1975. By 1986 half of Japanese outward investment was destined for the USA, induced by threats of protectionism (e.g. quotas on car imports), incentives by state governments jockeying for investment, and sharp appreciation of the yen. Europe follows, for similar reasons. The allure of a foothold in the European

Common Market of 1992 has attracted much Japanese manufacturing f.d.i., like Nissan to Sunderland, Toyota to Derby and Honda to Swindon.

Since 1986, however, more Japanese f.d.i. has been flowing to neighbouring East and South East Asian countries where, in 1987 for the first time, China displaced Hong Kong as the main host country for Japanese capital, followed in descending order of rank by South Korea, Indonesia and Singapore. This redirection reflects rising labour costs in the Gang of Four countries compared to China. Yet Japanese f.d.i. in the whole region, especially for making components and subcontracted consumer goods, has enabled Japanese multinational firms to maintain low production costs and market products aggressively in America and Europe. Overall, therefore, it is clear that the rise of the multinational enterprise, with the high capital mobility that it brings in the flows of capital, management, new technology and different production systems, has been a key feature of global economic change.

A CASE STUDY: THE CAR INDUSTRY

Car manufacture is a 'mature' activity, and one that illustrates well many of the influences at work in global changes in recent times. Two trends indicate the industry's maturity. First, growth in world car output and ownership, rapid in the 1950s and 1960s, dramatically slowed after 1970 (see Table 4.2). Though the slowdown was accentuated by oil price rises which inflated production costs, made cars expensive to run and reduced demand, it also reflected saturation of northern markets, where most sales replace existing cars. Second, making car parts and car assembly have become more standardised, requiring less skilled labour. Thus decentralisation of the industry from traditional locations has become a key feature since 1970. In this respect, car production mirrors more general global industrial shifts.

Decentralisation sharply altered the importance of producers in the North and made the South more prominent (Table 4.2). Former American dominance of world output has been replaced by a situation in which manufacturers in Western Europe, Japan and North America fight for a share of global markets. Significant shifts of car manufacturing have also occurred within Europe, where Italy, France, Belgium, Spain and the USSR have all displaced the UK as major car producers, while elsewhere the NICs have gained prominence. What forces underlie these changes?

Table 4.2 The changing world pattern of car production, 1960–88

Area/Main countries	1960 Million	%	1970 Million	%	1980 Million	%	1988 Million	%
'North'	12·66	99·0	22·2	97·6	27·0	92·8	31·1	93·7
North America	7·0	54·7	7·6	33·3	7·2	24·7	8·0	24·0
USA	6·7	52·1	6·6	29·2	6·4	21·9	7·1	21·4
Canada	0·3	2·6	1·0	4·1	0·8	2·8	0·9	2·6
West Europe	5·04	39·4	10·8	47·3	11·4	39·5	13·6	41·4
Belgium[1]	–	–	0·7	3·1	0·8	2·7	1·2	3·7
France	1·1	8·6	2·5	10·8	3·5	12·0	3·2	9·7
Italy	0·6	4·7	1·7	7·6	1·4	5·0	1·9	5·7
Spain	0·04	0·3	0·5	2·0	1·0	3·6	1·5	4·5
Sweden	0·1	0·8	0·3	1·1	0·3	1·0	0·4	1·3
UK	1·4	10·9	1·6	7·2	0·9	3·1	1·1	3·4
West Germany	1·8	14·1	3·5	15·5	3·5	12·1	4·3	13·1
East Europe	0·2	1·6	0·6	2·8	2·1	7·2	2·0	6·0
Poland	0·01	0·1	0·06	0·3	0·4	1·3	0·3	0·9
USSR	0·14	1·1	0·3	1·5	1·33	4·6	1·3	3·9
Yugoslavia	0·01	0·1	0·06	0·3	0·2	0·7	0·3	0·9
East Asia, Oceania	0·36	2·9	3·6	15·7	7·3	25·1	8·5	25·6
Australia	0·2	1·6	0·4	1·7	0·3	1·0	0·3	0·9
Japan	0·16	1·3	3·2	14·0	7·0	24·1	8·2	24·7
'South'	0·13	1·0	0·5	2·4	2·1	7·2	2·1	6·3
Brazil	0·05	0·3	0·25	1·1	0·7	2·4	0·7	2·1
Korea	–	–	0·01	0·0	0·2	0·7	0·8	2·4
Mexico	0·03	0·2	0·14	0·6	0·3	1·1	0·3	0·8
World total	12·8	100·0	22·7	100·0	29·1	100·0	33·2	100·0

Notes: [1] All but 0·2 million units from Belgium are wholly assembled from parts manufactured in other European countries. Statistical authorities assume, therefore, that such assembled units, which numbered 1 million in 1988, duplicate those in other European countries.
All figures are rounded.
Sources: *The Economist, The World in Figures* for data for 1960, 1970 and 1980; Society of Motor Manufacturers and Traders, *World Production Data 1988*, for 1988

Multinational firms are mainly responsible for the industry's growth and dispersion in the First and Third Worlds. North American output is controlled by the 'big three': General Motors (GM), Ford and Chrysler. Firms like Toyota, Nissan and Honda have spearheaded Japan's phenomenal rise, while most European cars come from firms like Volkswagen (VW), Renault, Peugeot–Citroen, Fiat and subsidiaries of the US corporations. These multinationals also brought car assembly to Third World countries. But, until the 1970s, they concentrated production in integrated plants. Ford's plants at River Rouge (Detroit) and Dagenham (London) took in iron ore and coal at one end, and turned out finished cars at the other. Volkswagen in Wolfsburg, Renault in Billancourt (Paris) or Fiat in Turin, and the Soviet ZIL works in Moscow, were all much the same. Today, improved transport has enabled firms to 'fragment' production among dispersed plants, each specialising in certain parts or in final assembly. But they have done so mainly in response to oscillating currency exchange rates which cause international fluctuations in the production costs they seek so hard to minimise. Ford Europe, for example, makes engines in Bridgend (Wales), gearboxes in Bordeaux (France), electronics in Cadiz (Spain) and assembles different types of cars in Spain, Belgium, England and Germany.

The 1970s oil crises pressured firms to search for ways to lower costs. They innovated fuel-efficient engines and small, light vehicles using alloys and plastics (instead of steel) to reduce car prices and running costs and so reinvigorate sales. As competition hotted up in stagnant markets, firms needed larger-scale, specialised and lower cost plants; so they relocated some output, mainly of component suppliers, to places where cheaper production factors – especially land and labour – could help them achieve profits from sales of small, low-priced cars. This explains why Ford, GM, Renault and VW have set up new small-car assembly plants in Spain and why Fiat made engines in Brazil and Poland for its 'minis'. Decline of the UK motor industry followed from poor management, increased labour costs, strikes, union resistance to new technology and work practices in the 1970s, and from high unemployment which squeezed home markets in the 1980s. All this facilitated rising import penetration by cheaper foreign-made cars and 'pushed' plant capacity abroad. The industry became a symbol of Britain's de-industrialisation but has been revitalised by Japanese investment.

Many 'pull' factors have also encouraged firms to locate production abroad. National policies have played a key role. Under UN encouragement in the 1960s, Third World governments fostered foreign investment in manufacturing to serve local markets protected by tariff walls. This attracted Fiat, Ford, GM, VW and Peugeot to establish plants in Latin America and the British Motor Corporation (later British Leyland, now Rover) to open production in India. But often the plants make outdated models for local use as they are protected from imports of new cars: Argentina still makes a Ford Falcon, Mexico the VW 'Beetle', India the Morris Oxford. Further global decentralisation of the industry has been augmented by state policies to introduce national manufacturing of cars for export. Recent sales launches in Western Europe of the Proton from Malaysia and Stellar from Hyundai of Korea are examples which follow in the 'treads' made earlier by the Soviet Lada, Polish Polonez, Czechoslovak Škoda and Yugoslav Yugo.

CONCLUSION: TOWARDS ONE WORLD

The preceding pages offer sweeping brush strokes on the canvas of the world economy. The broad frame seems remarkably stable: East–West division and dominance of North over South. Beneath, though, is kaleidoscopic change. The Pacific is seriously challenging the Atlantic as the 'workshop of the world'. Industry is dispersing from older centres in the USA, Europe and Japan into 'sunbelt' areas and NICs. Unfortunately, however, inflation, debt, political and military instability have limited further manufacturing diffusion into the middle and low income South. Many forces shape these trends. This chapter has focused mainly on technology, transport and communications and the spread of multinational firms, but the car industry briefly illustrates also the roles played by floating currency exchange rates, cheaper factors of production, and state policies to foster manufacturing to supply national or international markets.

Tightening, and more complex, interdependence has typified the industrial market economies since 1945. Trade in manufactures boomed within and between Western Europe, North America, Japan and Australasia, gradually intensifying competition between firms. For thirty years, raw materials and oil flowed from the Third to the First World in exchange for manufactures. Workers followed, in search of better paid jobs. That helped to dampen costs and thus sustain industrial growth in the northern market economies. Industrialisation in the Second World – which was largely isolated from the outside world – depended mainly on vast Soviet raw material and energy resources, and Polish coal, and on their barter for East European manufactures.

Interdependence changed after 1975. Recession halted migration: workers returned south and those staying in the North faced unemployment. Capital flowed more copiously to the NICs as multinational firms and local industrialists began tapping their advantages of low labour costs and substantial skill. As the NICs developed, they sold manufactures in exchange for services from northern market economies, where uncompetitive manufacturing industries declined. Materials and labour movement from South to North stagnated or shrank as cartels such as OPEC needed workers to expand local processing for export and as knowledge-based fifth wave functions began to displace older material processing and standardised labour-using industries in the North. OPEC members became focal points on the globe and from 1975 to 1985 labour poured into, then out of, Middle Eastern and African oil states. 'Petro-money', fuelling global finance, flowed from these countries into banks in London, New York, Frankfurt and Zurich and, from there, to Latin America, Asia, Eastern Europe and the USSR. Now there is a chance that, by 2000, demilitarisation and political and economic reform in the former Warsaw Pact countries of Eastern Europe will further open their economies to trade and so propel the Second World into closer interdependence with the First and Third Worlds, just as in the case of China since 1978. The global kaleidoscope of change whirls on.

QUESTIONS

1 Discuss the advantages and disadvantages of seeing the world as three, four or five 'regions'.
2 What have been the main factors leading to differences in the economic development of different regions since the 1960s?
3 Evaluate the relative importance of technological changes in transport and production in shaping global economic trends over time.
4 What factors have led to the growth of multinational enterprises (MNEs)?

FURTHER READING

Dicken, P. (1986) *Global Shift: Industrial Change in a Turbulent World* (Harper & Row: London). An excellent introduction.
Knox, P. and Agnew, J. (1989) *The Geography of the World Economy* (Edward Arnold: London). A good regional view of change.
Gwynne, R. N. (1990) *New Horizons? Third World Industrialisation in An International Framework* (Longman: London). A detailed examination of change in the Third World.

Case studies

Mounfield, P. R. (1984) 'Industrial UK up-to-date', *Geography* 69 (2), pp. 141–66.

Ramphal, S. S. (1985) 'A world turned upside down', *Geography* 70 (3), pp. 193–205.

Warren, K. (1985) 'World steel: change and crisis', *Geography* 70 (20), pp. 106–17.

5

NATIONAL PERSPECTIVES ON GLOBAL ECONOMIC CHANGE

Robert Bennett

THE IMPACT OF GLOBAL ECONOMIC CHANGE

The growing interdependence of national economies is an important
new feature of economic change. It creates many problems, among
which is the fact that individual nation states are becoming less fully
in control of events that affect their development and prosperity.
They seek to cope with such problems by a range of possible
strategies. But the roles that nations can play are affected by many
factors, including their relative size, their 'strategic' position, the
scale of their political and other power, their social and political
objectives, and the degree of 'openness' of their economies to the
influences of economic modernisation. Because of these, and other,
differences each country will develop its own national perspective
and have a different capacity to adapt to global change. As with
nations, so also with sub-national regions and cities within nations.
Regions and cities also have to respond to the impact of global
change.

This chapter addresses the problems faced by nations and regions
in adapting to change. Similar problems of adaptation have occurred
in the past, albeit under very different circumstances. The new
patterns of economic change require us to reappraise the relations
between people and the regions in which they live, to reassess the
role of governments and to analyse the new relationships of nations
to each other.

From such challenges arise new geographies of development.
These reflect the many tensions of adaptation to change and require
the creation of new roles for old areas. New economic and political
cores of development are emerging. For finance these are chiefly in
the very restricted centres of New York, London, Hong Kong and

Tokyo. For production the new cores are increasingly in the newly industrialising countries (NICs) of the Pacific Rim and Latin America, using methods of global integration of production through transnational corporations (TNCs). For management and political oversight, the role of nations is being eroded in favour of international agencies and associations – the EC, NATO, GATT, etc. For cities a different set of challenges awaits as, for example, the accommodation to continuing dispersal and decentralisation of population. For people, too, there are new opportunities to be derived from access to global information and global markets, and new difficulties to be faced as job creation and destruction take place more swiftly than ever before. In this chapter we address these problems through a discussion of the markets for capital and labour.

CAPITAL MARKETS AND NATIONAL GEOGRAPHIES

Capital is one of the three economic 'factors of production' which must be combined to produce any commodity or service. (The other two are the human resources of labour and skills, discussed later, and the land and its natural resources.) Capital is perhaps the most difficult of the factors to understand, since it is not always immediately visible. Indeed, the growth of information technology and rapid communications permits many capital movements to occur (such as stock and share transactions and international currency deals) without even an immediate piece of paper to show.

None the less, all capital ultimately consists of accumulated assets. Thus capital eventually appears as highly visible items such as factories, production plant and machinery, offices, infrastructure and exploited natural resources. There is also, however, an increasing stock of capital assets derived from intellectual research and endeavour, most notably in computer software and patents of new technologies.

The economic history of the world is wrapped up in the development of new ways of defining capital and in new means of transmitting it from one place to another. All past leaps forward in development have involved innovations in the definition and transmission of capital. At the root of capital, however, is always its capacity to earn income for its owner. A capital asset which cannot earn income has no value and therefore ceases to be an asset; instead it is neutral or even becomes a liability. The capacity of assets to earn income is at the root of economic development, and affects the relative wealth of their owners, and of the regions and nations in which they are located. Without capital invested in natural resources, factories and so forth – as well as the human capital of people – regions and nations have no assets and hence no potential to earn income.

Innovations and new technologies change the value of assets, particularly as new products and new means of producing them are developed. These changes in turn modify the value of capital, which, in its turn, influences the economic development and income levels of regions. The capacity of a machine or factory to earn income is rooted in the decisions on location of a particular period of time.

Geographers have observed a continuing series of changes in the location patterns of industry. These are related to the 'long waves' of development referred to in Chapter 4. Each wave of industrial development lays down a different geography in each time period. Each new geography depends on the locational requirements of that time. New technologies lead to new geographies and to the emergence of new 'core' regions, and to new 'peripheries' of areas left behind. Pressures grow for 'old' regions to be recast and to attempt regeneration by renewal of their machines, factories, infrastructure (and usually also their human skills). The special problems of such regional adaptation are outlined in Chapter 6.

A major recent change has been innovation in the way capital is organised and made more 'flexible'. Production methods and organisation, labour organisations and managerial practices have all been affected. Of all new developments, however, the most crucial has been technological change in computers, communications and robotics, which has allowed more dispersed production. Because of the leadership of Japan in this field some commentators refer to this phenomenon as 'Japanification'. However, the developments are more widespread than just the influence of Japan.

A crucial part of the changes implied by 'flexible' structures is shown in Figure 5.1. This contrasts traditional pyramidal structures of relations between firms with the emerging flexible forms. The pyramid was characteristic of firms in the 1960s. It characterised their relations with each other in that a small number of large firms had a few large suppliers and subcontractors, who in turn had their subcontractors, and so on. This created dependence both up and

Figure 5.1 The changing relations between firms: flexible (upper) and pyramid (lower) forms

FLEXIBLE FORMS

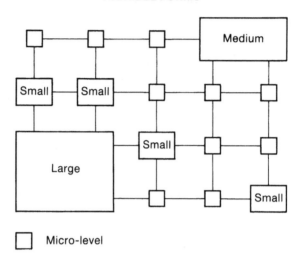

PYRAMID

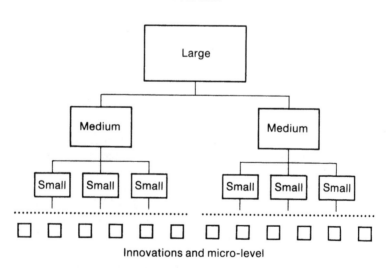

down the size spectrum. There was little independence, or opportunity to overcome the rigidities of supply relationships and permit ready adaptation to change. The capacity to innovate was restricted. The pyramid also normally characterised the structure within firms. There was usually an extensive network of hierarchical relations between managers and sub-managers: the so-called cost centre or line management approach popularised, amongst others, by management consultants such as Drucker and McKinsey.

Flexible structures, by contrast, do not have rigid relations between small and large firms but a changing and evolving network of contracting and subcontracting across each stage of production. This has been primarily associated with reduction in the size of direct employment by large firms. Even the world's largest global companies, such as BP or IBM, employ only 90,000–100,000 people. More commonly now, large companies contribute research, design, organisation and marketing. Manufacture is normally sub-contracted. As a result of these interlinkages, births and deaths of enterprises are easier. This allows more rapid innovation and entry of new ideas by readier creation of small firms, and faster and less painful removal of inefficient firms. The same structure is also beginning to characterise the internal structures of business management. There are now less rigid distinctions between managers and workers and greater incentives for individuals to perform in a wider variety of different environments. These will be your work experiences when you enter the labour force.

Flexible structures have also allowed the owners of capital more freedom to choose from a world selection of key locations. As a result, fewer locational decisions are now taken within the context of a single country. Instead, different regions have to compete directly with each other in global markets.

The emergence of Japan and the NICs of the Pacific Rim to prominence in the global economy has been based on the owners of capital choosing these locations, initially because of one dominant locational advantage – cheap labour. In earlier times the tech-nologies of production in these locations were often old and obsolete. But innovations in transport and communications made faster development possible, particularly by the use of containers, bulk carriers, air transport, telecommunications and satellites. Sub-sequently some of these countries, notably Japan and Korea, have been able to develop their research and development base to the extent that they can challenge, and even surpass, the traditional industrial countries in new technologies.

These changes have led to four sets of geographical developments in the 'Western' world. (1) Japan, and latterly also South Korea, have become leaders in R&D and organisational change in the structure of capital. (2) The older industrial countries of Europe and North America have had to adjust, first by suffering the hardships of the decline of older manufacturing, and then by grasping the opportunities for change by adapting industries and developing new organisation structures and investing in new R&D. (3) The newly industrialising countries of Korea, Taiwan, Singapore and Hong Kong (and latterly Thailand) have depended largely on cheap labour for export penetration of the markets of the industrialised countries. They now have the challenge of trying to follow Japan in developing local bases of R&D to offset the cost-saving developments now being applied in the old industrialised countries. Only Korea so far seems to have made major steps along this road. (4) The poorer developing countries have been left behind, seeking to emulate the NICs, but with major barriers created by poor infrastructure, often extreme pressures of population numbers, frequently major environmental hazards (of disease, drought and famine), and an insufficiency of capital resources.

The key challenge to nations is, therefore, to promote capital investment in various ways. First, to lay down the infrastructure of roads, ports and communications which permit modern trade links to develop. Second, to utilise local labour skills through investment in industries for both export and home markets. Third, to build on the resources of local capital to develop a national R&D capacity. Innovation, R&D and human adaptation to change are, therefore, the key aspect determining present and future capital investment. This leads us next to analyse the human element in industrial production.

HUMAN RESOURCES AND NATIONAL GEOGRAPHIES

Like capital, the human resources that produce a commodity or service can take a variety of forms. These range from the 'brute force' of human energy, such as digging resources from the ground, through jobs requiring various levels of specialist skill such as machining wood, metal and other materials, to skills of a more intellectual character such as design, presentation, manipulation of documents, or innovation of new software and products. The tendency has been for each technological revolution in production (the 'long waves' discussed in Chapter 4) to increase the level of specialist skills and intellectual inputs required, and to reduce the human energy inputs. The most recent technological revolution has increased the need for intellectual skills even further. This has tended to shift labour market demands away from manual and skilled trades towards a much higher level of intellectual skills and training, whilst at the same time shifting the industrial mix from primary to secondary, and increasingly to tertiary activities (see Chapter 6).

National labour markets, particularly in the older industrialised countries, are therefore experiencing a range of important challenges. How can their present labour resources be best used to maintain and increase incomes and standard of living? For such countries this is requiring re-skilling, improved labour responsiveness to change and the anticipation of new needs.

Such challenges are not easy for any of us. Indeed, a recent report by the Training Agency in Britain referred to the problem as a 'challenge to complacency'. This report, and other recent studies, have recognised the appalling record of Britain in education and training and its apparent inability to provide the human resources needed by the changing economy. The blame for this is widely shared. Industry has taken insufficient interest in training its workforce, often preferring the short term approach of 'poaching' trained personnel from neighbouring firms with good training policies. School education has also often been inappropriate. One-half of school leavers in 1979 had no examination qualifications whatever, while the qualifications of those that did frequently had insufficient relevance to the world of work.

Many reforms are now taking place, and training agencies can help to fill the gap. But here again Britain has many problems. Such agencies, it is argued, have been dominated by approaches geared to the needs of older industries which are in decline, while some government training schemes have been conceived of more with the aim of getting people off the unemployment register than with providing genuine skill training. The trade unions, also, have been highly resistant to change, frequently blocking training schemes, and taking the short term view of defending traditional jobs and skills rather than promoting new skills for the new jobs becoming available.

ROBERT BENNETT

While this description is a caricature, it is not so unjust for the case of Britain, which perhaps represents the extreme among the advanced nations in neglecting the skills of its population. For the industrialised countries as a whole, however, the problem of adaptation to change is a general one, affecting everyone – management, educators and unions alike. During a revolution in the technologies of production it is inevitable that existing approaches to labour development will be inappropriate and will have to change. Resistance and political controversy are almost bound to arise.

Some aspects of the problem are evident from Table 5.1. This demonstrates the poor position of Britain in the education of its population after the age of sixteen. The USA, Germany and Japan provide marked contrasts, with over three-quarters of their populations having post-sixteen education. Britain's higher education record is also very poor. However, perhaps the most important issue is the skills of the school leaver. Here the British record is particularly weak. In 1979 only 50 per cent of UK school leavers had any formal qualification, compared to over 60 per cent in Germany, Japan and the USA. Major initiatives have recently been targeted at this problem and some improvements have taken place. Let us look briefly at four national cases in more detail to see the nature of responses that are possible.

Germany has approached its labour market needs with probably the best method of vocational and skills training in Europe.

Table 5.1 Education and training experience in four countries, 1979–81

Experience	Germany	USA	Japan	UK
Total participation in education and training 16–18	84	79	73	63
Total in full time school education 16–18	31	69	58	18
Workers with recognised qualifications equivalent to at least one CSE pass	66	78	60	50
Per cent of population gaining higher education qualification (at all levels)	20	32	37	18
Per cent of labour force with university degree	8	19	13	7

Sources: National Economic Development Office, *Competence and Competition*, 1984 and DES Statistical Bulletin 10/85

110

Approximately 70 per cent of its population enter apprenticeships at the age of fifteen-plus and 20 per cent go on to higher education. The training groups are supported by what is called the 'dual system'. Through this, employers provide the main training in jobs and bear 80 per cent of the cost, whilst the government (through the states, or *Länder*) provide 20 per cent of the costs and fund day release colleges (*Berufsschulen*). The training content of the college courses and of the firms is monitored by a collaborative arrangement between employers and unions and implemented through local Chambers of Commerce. The result is highly impressive: there are over 450,000 'approved training firms' with an annual intake of 600,000 apprentices. The success of the German approach, however, depends not only upon its size. Most crucial is the consensus it represents between business, government and trade unions. The core of this consensus is the aim of securing the conditions for the economic success of business as the chief guarantee of the success of the nation. Central to this aim is the objective of giving every person entering the labour market some occupational competence and qualification. The key aspect of the national approach to economic change is for Germany to compete with the rest of the world through greater occupational competence, which assures a high quality and adaptable workforce. Moreover, Germany has been most successful in developing competence in traditional sectors (such as engineering and chemicals), but through radical re-skilling and high-technology investment.

The USA is large and diverse, and fifty individual state governments have educational responsibility. This leads to great diversity in educational provision. However, there is one key cultural characteristic: that economic advantage is seen as a justifiable reward of personal commitment to a job. This has led to widespread support for a range of school and further education options which, although not perfect and subject to many criticisms and reforms to improve quality in the 1980s, offer probably the highest level of participation in any industrialised country. From high school 71 per cent of children obtain a diploma qualification, and most go on to training in employment. The scale is huge: there were 11 million Americans in employment training in 1981. Major American firms expect to provide extensive training. Leading companies like IBM, Xerox or Boeing spend 2·5 to 3·5 per cent of their sales revenue on training, and many firms have their own training schools. The trade unions are also generally supportive, both financially and by entering into training trusts with employers. Training is seen by almost all as a key means by which the US economy can stay abreast of its international competitors.

Japan is characterised by a cultural 'thirst' for education, with 94 per cent of the population staying on after compulsory education ends at the age of fifteen, and 37 per cent going on to university. Education is an important part of labour training, but not just by providing vocational skills. Rather, the emphasis is on a high level of general competence (especially literary and mathematical), as well as socialisation and harmony with labour market needs. As a consequence the average age of people entering jobs is twenty. In work, there is a perspective of 'lifetime education'. This encourages adults to attend courses in their own time. Employers offer extensive and continuous on-job training, often in their own schools, which can attain the size of universities. When skill needs change, employers provide the new training. All this is supported by state subsidies and manpower strategies which have sought to identify new skill needs and opportunities ahead of time. The result is an amazing level of organisational flexibility in both labour and management which has given the Japanese economy the leading edge in its ability to respond to, and indeed to lead, change in the world economy.

Britain, by contrast, has been a laggard in many respects in its response to changing labour market needs. But it is easier to outline the areas which need to change than to achieve such changes. Probably the key needs are (1) a major increase in the involvement of business in schools and training schemes, (2) a massive increase in the number of people staying on in schools or colleges after sixteen to obtain greater skills and qualifications, (3) a development of the methods of assessment and gaining qualifications in schools to give all school leavers a recognised qualification, e.g. through records of achievement, (4) youth and further employment training schemes emphasising occupational competence rather than loose 'training for stock', and (5), most difficult of all, a change towards a greater general awareness of the need for skills and competence by firms, individuals and institutions such as trade unions. Short term advantage and narrow interests need to be replaced by the longer term emphasis on the skills necessary for the continued success, competitiveness and expansion of the British economy.

A major stimulus to these necessary changes has been given by the rapid development of closer links between business and education. One major initiative has been the drawing up of compacts between schools and colleges and local employers. In addition to compacts, however, many educational institutions have developed close partnerships with business. These involve work experience, or curriculum projects and work shadowing by pupils and teachers. In another case, the London School of Economics offers an example of leadership in attempts to forge higher education links with schools in London to encourage greater staying on after sixteen. These developments are summarised in Figure 5.2. This figure, based on research undertaken at LSE, for the Foundation for Education–Business Partnerships, shows how such partnerships have rapidly developed to include greater depth in an increasing number of education authorities. The objective for the future has to be further movement of partnerships beyond level 3 (which is that of a compact) to levels 4 and above, where all schools and all year groups are involved in detailed curriculum projects

Figure 5.2 Partnerships between education and business, 1986–90. How local education authorities in Britain are forging closer links between schools and business. At level 1 there is little or no connection; level 2 represents various *ad hoc* links; level 4, deep curriculum involvement. Vertical axis represents the proporation of all local education authorities in each category.

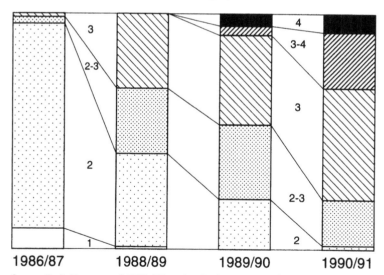

Source: R. J. Bennett and CBI, *Education–Business Partnerships: The Learning So Far*, CBI, London, 1992

with, or about, business. This development offers the hope of equalling the skills developed in Japan or Germany. Only such close linking between education and the needs of the workplace will assure Britain's capacity to compete effectively in the future.

REGIONAL PERSPECTIVES

The global context of change has required new national response strategies to be developed. A key aspect of current debates has been a shift in thinking by managers and government away from initiatives 'from above' in favour of development 'from below'. Nations have had to grapple with situations in which their various regions often have very inadequate capacity to respond to change. Traditionally, initiatives 'from above' have sought to use the government stimulus of regional policy or urban policy. This particularly characterised Britain in the 1960s and early 1970s. The alternative of development 'from below' seeks to address local needs and remove the barriers to local initiatives for the development of infrastructure and labour skills and the improvement of local environments.

The change of approach that has occurred has sought to identify, and provide, the conditions for regional economic success. Nowhere has this question been more focused than in the case of 'high-tech' and new technology industries associated with computing, micro-electronics, biotechnology and pharmaceuticals. But the same questions are now being addressed also for a wider bundle of industries ranging from the basic manufacturing of vehicle components and clothing to delivery of services in the areas of finance, law, and governmental functions (such as personal care, health, refuse collection, and the like).

Associated with this search for the regional conditions necessary for economic success has been the fact of growing competition in a global economy. Success in such competition depends upon getting 'supply side' factors right. Chief amongst these factors are the right capital and labour markets: a good labour force with the right skills; a financial base to support infrastructure and initial development by sharing risk; an entrepreneurial and management capacity to organise; sufficient technological and information links to create a 'critical mass' of activity within a local context. The relation of these factors to social environmental conditions is shown in Figure 5.3. As a result of this emphasis, regions and localities have increasingly

Figure 5.3 Local conditions of successful development. How local conditions relate to the key actors that affect them.

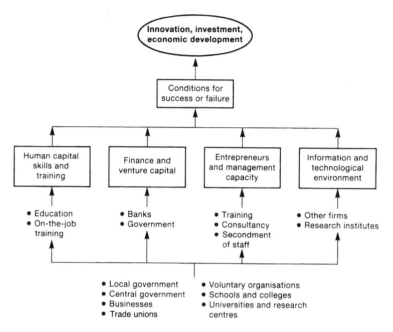

joined in a 'game', competing with each other for new industrial development. Indeed, the local development game has increasingly replaced, or at least supplemented, national perspectives in the competition for development.

Some commentators question the value of regions and localities competing with each other for mobile investment, either within one country or between countries. They argue that the outcome will not be an increase in the overall level of economic prosperity – but merely a geographical transfer of firms from one place to another. This criticism must be correct if geographical relocations are the sole outcome of local economic development. But there are two counters to this criticism.

First, if the development conditions can be got right in each place, this should increase the level of genuinely new economic development, and thus contribute to an improvement in incomes and wealth as a whole. Some commentators refer to such a process as developing local 'synergy'. This means that the combination of a number of local factors, under the best conditions, will lead to more growth than would be possible if those factors acted independently or in different places. This provides a basis for local, or *endogenous* development of an area. Synergy thus means that the overall level of economic prosperity will increase by people working together in a form of 'partnership'.

Second, there has been a growing understanding that economic success must be founded on local leadership. Endogenous development is not a spontaneous process. Frequently more forces work to pull people and agencies apart at local level than draw them together. The advantage of synergy will be gained only by a sustained effort to draw efforts together. Local leadership is crucial to this process. If the key local 'managers' can work together and share a vision or strategy for an area, this should allow others to follow and to understand where their contributions lie. This means local education leaders, especially head teachers, acting together with local businesses, colleges, universities, community groups and business support services such as Chambers of Commerce, enterprise agencies or Training and Enterprise Councils (Local Enterprise Companies in Scotland).

These arguments mean that the old explanations of economic geography based on transport costs and the regional comparative advantage of different locational factors are no longer so important, because of the declining relative significance of transport and raw material costs compared to the costs of labour and capital. Today's understanding of geography perceives the importance of close local co-operative links between capital and labour. This involves co-ordination and collaboration between local people and local agencies to provide the local capacity for a region or locality to act successfully in the global competition.

The shapes of the new local geographies are only now forming and we can only speculate how the future will develop. In the next chapter we look at some case studies of economic change in regions in order to see what shape the geographies of the future may take.

QUESTIONS

1 What capital resources are needed for the economic development of countries?
2 What do you understand by the 'flexibility' of production methods?

3 How do demands on people's skills differ now from those in the past?

4 How do British education and training differ from those in other countries and how adequate are they?

5 In what ways does 'development from below' differ from traditional 'regional policy'?

FURTHER READING

Dicken, P. (1986) *Global Shift: Industrial Change in a Turbulent World* (Harper & Row: London) especially chapters 1, 2 and 3. This is an excellent introduction to changes in the global economy. Later chapters give detailed accounts of major sectors and types of firms.

Hansbury, J. (1987) *Secondary Education and Jobs: A Guide for Parents* (Merlin Books: Branston). Useful to teachers, parents and pupils. In thirty-eight pages it presents a view of English education and the need to introduce a greater emphasis on skills and competence.

Warwick, D. (ed.) (1989) *Linking Schools and Industry* (Blackwell: Oxford). A set of short essays giving overviews and case studies of compacts, enterprise and curriculum development in schools. Useful for teachers and in student project work.

Bennett, R. J. and McCoshan, A. (1993) *Enterprise and Human Resource Development: Local Capacity Building* (Paul Chapman Publishers: London). Gives detailed coverage of British education and training developments linked to local economic development.

6

NEW ROLES FOR OLD REGIONS

Robert Estall

STAGES OF ECONOMIC DEVELOPMENT

The powerful forces of change currently at work in all parts of the world have greatly affected the geography of economic activity. Patterns of economic life have been altered and relationships between areas transformed. The fortunes of different regions have changed, often for better, sometimes alas for worse. This chapter examines some of the experiences of change in 'old' regions. 'Old' regions could be variously defined, but here their identification is based on comparative levels of economic development in modern times, and the terms 'advanced' or 'mature' could be used interchangeably with 'old'.

Economic development generally proceeds through a series of stages, in each of which certain kinds of productive activity achieve prominence. This process may be measured by using three broad categories (sectors) of activity: primary, secondary and tertiary. Early stages of economic growth (i.e. in 'young' regions) normally find most of the working population in primary activities, especially producing food or basic material requirements like fibre and lumber and some minerals. With the introduction of ideas that increase primary productivity, however, surpluses are created for sale. Incomes rise and more people are able to become engaged in the secondary occupations of manufacturing and construction (chiefly in manufacturing industry) which gradually become the key wealth-producing activity. But manufacturing must be supported (as must improved farming) by better services, especially at first the transport and trading functions. Thus the share of jobs in the primary sector steadily declines, while that in manufacturing and related services rises.

As labour-saving technologies develop further, productivity in goods production (both primary and secondary) continues to improve. This creates more wealth, and it becomes possible for people to allocate an increasing share of their purchasing power to personal and other services. The services demanded by the goods-producing industries themselves also become more numerous, as in advertising, legal and financial services. As economies advance, therefore, a steadily rising share of the working population finds employment in the tertiary sector. These long-term trends are portrayed in a highly generalised way in Figure 6.1, which reflects what is known as the 'sector theory' of growth.

Such change in employment is reflected also, of course, in the contribution made by the different kinds of economic activity to a region's income. At a global level, we can conveniently regard whole countries as 'regions', and data available for the years around 1980 (Table 6.1) demonstrate the contrasts between old regions (like the UK, USA and West Germany), newly industrialising areas (like Mexico, Greece and Brazil) and poor Third World economies (like Bangladesh, Ethiopia and India). The three major sectors of activity have very different shares in the gross domestic product (GDP) of these countries, and this fact is directly reflected in the income of their populations.

Figure 6.1 Shares of jobs at different stages of development

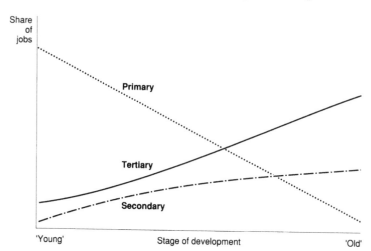

The economist Colin Clark, an early worker on sector theory, found a 'firmly established relationship' between a region's economic structure and its wealth. High levels of real income per head, he argued, were always associated with a high proportion of the working population engaged in tertiary activity. Low incomes were always associated with a high proportion in the primary. Not everyone accepts the certainty that Clark suggests, but the broad relationship is obvious. The final column of Table 6.1 gives the value of the gross national product per head in each of our sample economies in 1980. The associations that Clark suggests are clearly illustrated. (Note: GDP is the value of all goods and services produced within the territory concerned, while GNP includes international transactions.)

Table 6.1 Gross domestic product by kind of economic activity, c. 1980 (percentage of total)

Country	Primary	Secondary	Tertiary	GNP (US$ per capita)
USA	3	35	62	11,470
West Germany	2	48	50	8,800
UK	2	35	63	7,950
Greece	15	28	57	4,690
Mexico	10	38	52	2,530
Brazil	11	29	60	2,030
India	33	23	64	230
Ethiopia	46	15	39	150
Bangladesh	54	14	32	130

Note: These shares are calculated from United Nations data in which 'Primary' includes agriculture, forestry, hunting, fishing, 'Secondary' includes the manufacturing, construction, mining, electricity, gas and water industries, and 'Tertiary' includes all remaining activities. Elsewhere in this chapter mining is included in the primary sector, while the electricity, gas and water utilities are included in the tertiary.

In sector theory terms, then, an 'old' region can be considered as one with a relatively long history of industrial development. Agriculture and other primary industries have been displaced from positions of prominence in employment and output and the role of manufacturing and, increasingly, services has advanced. The key to this development process lies in *productivity*, a point which this discussion will several times stress. Large and continuing advances in productivity have been made in old regions; first in agriculture (where the very small shares of employment and income conceal an enormous increase in volume of farm output), and subsequently in manufacturing. Such advances created much new wealth, so that our old regions are materially rich, as Table 6.1 suggests.

That does not mean that they face no problems. Their advance to maturity has not been trouble-free, and further progress in a competitive world demands further change. Old regions must continue to find new roles as younger regions challenge or usurp their position in older established activities. This is very evident today in the case of regions with high concentrations of manufacturing industry. As individual industries mature and need mass markets for large volumes of standardised products, low production costs become important for success. Labour costs, especially, can become more critical. But old industrial regions are often at a disadvantage here compared with newly industrialising areas. Long established industries in old regions tend to decline, therefore, and continuing prosperity requires structural, and spatial, change. This can bring pain for people and problems for places. High unemployment and localised economic distress occur, together with poverty and deprivation for some of the population. Such conditions may not often bear comparison with those of poor people and poor places in the Third World environments discussed in Chapters 8 and 9; but they are real enough to those concerned, and merit serious consideration.

GROWTH AND CHANGE IN THE USA

The United States of America provides a good example of the development stages through which regions pass. The processes of change are far advanced and, within such a large country, various regions can be at different stages of development. This has allowed many writers to refer, for example, to the sharp contrasts between the northern and southern regions of the USA (between 'snowbelt' and 'sunbelt'), which are discussed below.

This chapter is chiefly concerned with the challenges of recent change. But Figure 6.2 puts present patterns into historical perspective by illustrating the transformation of the American employment structure since the beginning of this century. The primary sector remained dominant around 1900, but subsequently lost ground quite rapidly. Today it is a very small employer. The most important changes in primary employment were those on the farm. Major advances in technology brought an absolute, as well as relative, fall in jobs. About 11 million people were employed on

Figure 6.2 The distribution of employment, by economic sector, United States, 1899–1987

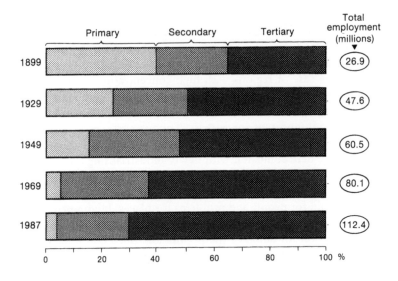

American farms in 1900, and still over 9 million in 1940. A great agrarian revolution then followed. This multiplied farm output but led to a collapse of employment to not much above 3 million by the late 1980s.

This record may be used to illustrate the problems which face old regions as their economies evolve. The revolution in American farming certainly brought success to many, and provided the bedrock on which the industrial and service sectors could advance and national living standards be raised. But it also brought hardships. Many dispossessed farm families were forced to leave their homes and move to cities. One estimate suggests that about 55 million people moved from farms to cities between 1933 and 1970. This huge migration (captured in literature by the great Steinbeck novel of 1939, *The Grapes of Wrath*) caused problems for both rural and urban communities. Rural towns, losing both people and enterprise, became less able to serve the remaining farm and rural populations. Many of the people left behind struggled on in increasing difficulty, and the incidence of poverty among the American farm population remains well above that for the population generally. Urban areas, for their part – and chiefly the larger cities which were favoured by the migrants because of the expected opportunities – were often unable to cope adequately with the flood. Rural people were generally unfamiliar with urban ways of life and many possessed few relevant job skills. This helped to create the very poor conditions that still exist in central parts of American cities. Clearly, although the nation as a whole benefited greatly from the changing structure of the economy, the adjustment to new roles was not easy.

Meanwhile the secondary sector (chiefly manufacturing industry) increased its share of jobs until after the Second World War (Figure 6.2). In fact, the peak share for this sector (about one-third of all jobs) came in 1953, when the demands of the Korean War boosted manufacturing employment. Since then the share has been substantially reduced, while in the 1980s the absolute numbers employed in manufacturing also tended to decline. This experience has given rise to suggestions that a 'de-industrialising' process is under way. In fact, however, the value of manufacturing output has continued to increase. For the period 1980 to 1987, for example, when employment in manufacturing fell by about 5 per cent, the value of output, in constant 1982 dollar values, rose by 26 per cent. (Constant dollar values need to be used in such calculations to allow for changing price levels and inflation.) The experiences of manufacturing thus begin to resemble those, historically, of farming, reflecting that basic mechanism of economic growth and change, rising productivity.

Finally, in the pattern of change that Figure 6.2 displays, the tertiary sector is seen to go from strength to strength, greatly increasing both its number and its share of jobs. Between 1970 and 1987 the American economy added an amazing total of about 30 million new jobs to its service trades; something like 1·8 million new jobs every year.

In summarising the long term changes shown in Figure 6.2, it is useful to refer to the combined primary and secondary groups as 'goods producers', as opposed to the 'service producers' of the tertiary. In 1900 roughly two out of every three jobs were directly involved in goods production, and only about one in three in the service trades. Since then, many workers have assumed new roles and the balance has been reversed. Today seven out of ten jobs are in the tertiary sector. Goods production has yielded pride of place to services, which dominate employment; yet Americans have become steadily wealthier, lending support to the propositions of sector theory.

CHANGE WITHIN THE MANUFACTURING SECTOR

We have seen that economic progress requires structural change. But as more and more workers in old regions are employed in services, and manufacturing declines relatively (or even absolutely in some areas), questions arise about the eventual consequences. How will the old regions fare? Some people argue that goods production and export are at the root of a region's prosperity. The erosion of its industrial base is thus a matter of great concern. Others believe that an efficient service sector can sustain a wealthy region, and that exports of services can replace exports of manufactured goods or primary products. A better understanding of these matters requires a look at changes taking place within the major sectors.

The various kinds of manufacturing industry have very different recent histories. This is shown in Table 6.2, which gives employment data for selected manufacturing industries in the USA for the period 1970 to 1987. (The table also gives employment projections for the year 2000 that assume a 'moderate' rate of national economic

Table 6.2 Manufacturing employment, by selected industry groups, United States, 1970–87, and projections, 2000 (000s)

Industry	1970	1980	1987	% change 1970–87	Projection 2000
A					
Textile Mill Products	974	847	725	−26	607
Leather and					
Leather Products	320	233	144	−55	98
Apparel and other					
Textile Products	1,364	1,264	1,100	−19	924
Food and Kindred Products	1,786	1,708	1,624	−10	1,456
Primary Metal	1,260	1,142	749	−40	574
B					
Electrical and					
Electronics Equipment[1]	2,065	2,445	2,576	+25	2,631
Transportation					
Equipment	1,852	1,900	2,048	+11	1,697
Instruments and					
Related Products	527	712	696	+32	771
Printing and Publishing	1,104	1,252	1,507	+37	1,706
All Manufacturing total	19,367	20,286	19,065	−2	18,160

Note: [1] Including electronic computing equipment, which is classified with 'Machinery except electrical' in the US Standard Industrial Classification.

growth.) The selected industry groups broadly represent the two extremes of recent experience, and together they accounted for 58 per cent of all manufacturing jobs in 1987.

Group A contains some of the staple industries of an industrial economy in its earlier stages of growth. They often show direct links with the primary sector, using the products of farm, forest or mine for their major inputs. Understandably, then, they would have accounted for a large share of industrial employment and output in old regions in the late nineteenth and earlier twentieth centuries. The available records suggest that the combined food, textile, apparel, and shoe and leather product industries still employed over one-third of all manufacturing workers in the USA in 1921. By 1987 their share had fallen to under one-fifth.

More significant, however, is that this long relative decline recently became an absolute one, as Table 6.2 shows. In the textile, leather, clothing and food industries the employment run-down began before 1970 but, except in textile mills (where employment peaked at 1·3 million in 1948), the losses had been small. But from 1970 the declines became sharper, and the future looks grim (Table 6.2, projections). In the very important primary metal group, too, employment was quite stable for many years before 1970, but this industry then fell upon disastrous times. It lost 40 per cent of its jobs between 1970 and 1987, and yet more losses are expected.

Such changes were probably inevitable. As the economy matured, and new products and processes were introduced, the long established branches of manufacturing were bound to lose ground relatively. In addition, such industries could also be expected to figure prominently in the early industrial development of other (young) regions of the world, where production costs (especially labour costs) were lower. Given free trade, therefore, and the global changes discussed in Chapter 4, employment in these industries in America was bound to be affected by import competition.

Much criticism has been levelled at both management and labour in these old staple trades for not ensuring that American firms remained more competitive via new investment and innovation. Such criticism may be well deserved. But it must be recognised that the introduction of more highly productive methods would itself normally be reflected in declining labour inputs. For example, while the decline of American primary metal employment in the 1970s was largely due to high costs and poor competitiveness, a good part of the continuing decline in the 1980s was due to greater investment in advanced production technologies. This improved efficiency, but at the price of many more lost jobs. Thus, whatever view one takes of the causes of the decline of such industries in old regions, their great days as employers are over, even if their output can be maintained.

When looking for new roles for old regions, however, it is not only to services that one needs to turn. The industries listed in group B in Table 6.2 illustrate some of the areas of opportunity in manufacturing – the so-called 'sunrise' industries. Here there are some relative newcomers to the industrial scene, like computers and advanced communications equipment. Others, like printing and publishing, are old industries that have been powerfully stimulated by the demands of modern life and the introduction of new technologies. The transport equipment industry is another such example. Its roots are in the ship, carriage and locomotive building of bygone ages. It still possesses a massive interest in the 'mature' motor vehicle industry. But the most significant area of the transport group's recent growth has been in the high-technology 'aerospace' fields – aircraft, guided missiles and space vehicles.

Similar claims to 'high-technology' status can be made for the other industries listed in group B of Table 6.2. The electrical and electronic equipment group, for example, has become a giant interest of the advanced economy. Not only does it produce a multitude of 'hi-tech' products for final consumers, but it also has a critical role in producing equipment for use in other manufacturing, and service, trades. The instrument industry, for its part, although relatively small in employment size, has had a very high rate of growth. This is another industry, rooted in advanced technology, whose products support developments in many other fields where measuring and control devices perform critical functions.

The past role of these four industry groups cannot be accurately measured because they contain so many products that simply did not exist before the Second World War. In 1947, however, having been greatly stimulated by wartime spending, they together accounted for about 19 per cent of all manufacturing jobs. By 1987 they accounted for 36 per cent. There is no single list of factors behind this positive experience. The rejuvenation of old industrial regions will depend on many influences, whose strength will vary from time to time and place to place. However, some key features can be highlighted.

A prominent factor in recent regional growth has been the

126

growing role of industrial research and development. But this, in turn, requires large financial resources, so that a very important feature has been heavy and consistent support from the federal government. Government interest in military and space technology and hardware has funded massive R&D programmes in the aerospace, electronics, computing and instrumentation fields, some of the results of which have subsequently found civilian applications. In many years federal money has met about half of total industrial R&D spending in the USA. Most of the rest comes from industry itself, but this private money has also been concentrated in the group B industries of Table 6.2. Thus R&D work is one of the most important keys to the creation of new roles. In America the number of scientists and engineers employed directly by manufacturing firms has grown by more than half since 1970, outstripping the growth of most other occupational groups. World famous concentrations of such workers have grown in places like Silicon Valley, south of San Francisco in California, and along Route 128 around Boston in Massachusetts.

Another important feature of the regeneration process is the quality of people and, especially, the availability of sufficient numbers possessing technical and managerial skills. American universities and colleges have been active in meeting these manpower needs. These institutions have also often involved themselves directly in the promotion of new high-technology enterprises, encouraging their staff to undertake outside work, and perhaps to start up their own firms. This, in turn, suggests the importance of the entrepreneur in the creation of new roles. Such business leaders are crucial for their practical role in applying new ideas and bringing new products to the market place. Someone must perceive opportunities, take the risks involved in establishing a new firm or introducing a new product or process, and oversee the critical early stages of development. This entrepreneurial capacity is among the most vital of all elements.

But entrepreneurs themselves depend upon the availability of investment capital, and this is another matter in which mature regions may have advantages. Such regions normally have sound and well established banking and financial structures, and command large financial resources. Admittedly, mobile capital from old regions has often been active in developing industry in other regions and countries. But funds will also be readily available for promising internal developments, so that the proximity of capital (especially for the risky early stages of new ventures) is another advantage that old regions may possess.

In sum, old industrial regions should not be simply shedding inherited manufacturing interests but also creating new points of growth. Even so, such new growth will not necessarily replace lost jobs in equal numbers. In the USA, manufacturing employment seems now to have passed its peak and may, if the projections offered in Table 6.2 are a reliable guide, be set upon slow long term decline. The key employment issue for the future is, therefore, the role of the service sector.

CHANGE WITHIN THE SERVICE SECTOR

The recent expansion of market demand in the United States has been immense. In part this has been a response to increases in population, but the chief cause has been a rapid increase in spending power. Between 1970 and 1987, for example, the American population grew by nearly 19 per cent (from 205 million to 243 million) but GNP, in constant 1982 dollar values, rose three times faster. This increase in market size and wealth brought a great surge in demand for services. This was faithfully reflected in employment, which grew by 64 per cent between 1970 and 1987. Not all kinds of services had such rapid growth, of course. Slower than average expansion occurred in the transport, communications and public utilities groups and in wholesale trade. These are large employers, but their workforce has recently grown more slowly than that of other services. This reflects their important role during earlier stages of economic growth. Their prominent place in the economy has been long established, and further great expansion of their share of jobs was not to be expected. (Figure 6.3, which subdivides the service sector into two groups by share of GNP, also illustrates this point. The 'A' services of transport, communications, utilities and trade provided 26 per cent of GNP in both 1960 and 1987.)

For other kinds of services, however, rising prosperity has brought a veritable explosion of jobs. Table 6.3 identifies activities where employment grew, between 1970 and 1987, by more than the average (about 64 per cent) for services as a whole. (Government employment is also included, however, because of its great size.) *Consumer services* in general have been powerfully stimulated. Within the retail trade sector, for example, employment in 'eating and drinking places' more than doubled between 1970 and 1987,

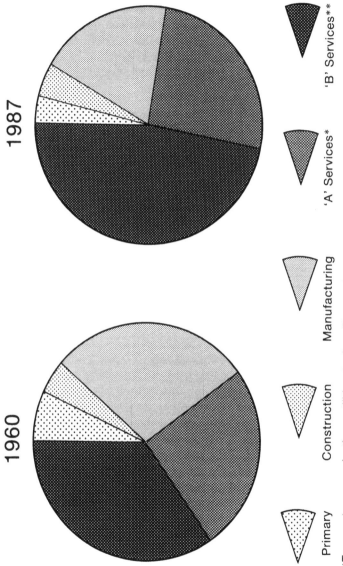

Figure 6.3 Gross national product, by economic sector, United States, 1960 and 1987

1960   1987

Primary   Construction   Manufacturing   'A' Services*   'B' Services**

*Transport, communications, utilities, trade **Finance, insurance & real estate, government, all other services

and has reached above 6 million in total. This single branch of retailing is now several times larger than the largest manufacturing industry. Wealthy individuals also spend more – much more – on health services, and they tend to sue each other more often, and generate many other legal needs, if the explosion of jobs in legal services is any guide! Increasing wealth also brings a demand for more public services, in defence, education, public health, the police and fire services, for example. This is reflected in the enormous numbers employed by federal, state and local governments.

Table 6.3 Tertiary employment, by selected service groups, United States, 1970–87 and projections, 2000 (000s)

Industry	1970	1980	1987	% change 1970–87	Projection 2000
Retail trade	11,048	15,035	18,509	+68	22,702
Finance, insurance and real estate	3,646	5,159	6,549	+80	7,917
Business services[1]	1,676	3,092	5,172	+209	8,121
Legal services	236	498	797	+238	1,267
Health services	3,053	5,278	6,828	+124	9,774
Government	12,553	16,241	17,015	+36	18,329
Tertiary sector total	47,147	64,384	77,526	+64	94,478

Note: [1] Includes computer and data processing, research, advertising, management, consultancy, personnel supply services, etc.

But just as important as such final consumer demand is the fact that many of these services are also required by business enterprises and by the goods-producing sectors. Efficiency in the production and marketing of goods has depended increasingly on the specialist supply of technical, financial, insurance, business, legal, trade and other services. These are known as the *producer services* to distinguish them from the 'consumer services' mentioned above. Much service employment is thus directly linked with the farms, forests, mines and factories of the goods-producing sector. Those who buy goods also buy, indirectly, the growing array of services that go into their production and marketing. Because so many services are purchased both by consumers and by producers, it is difficult to calculate the shares of jobs that depend on each. Probably about half of all service jobs are linked directly with goods production. But the other (consumer) half also depends heavily on income generated in the goods-producing sectors. It is in this context that the debate about the 'de-industrialisation' of old regions becomes important.

DE-INDUSTRIALISATION

Unfortunately, there is no agreed definition of the term 'de-industrialisation'. Broadly, it refers to the kind of change represented in Figure 6.2, where the manufacturing sector appears to be in decline. A widely held view is that, since goods production is the key to wealth creation, a large scale loss of manufacturing in old regions puts them in great danger. This is because much service employment is linked to goods production, either directly, in the producer services, or indirectly, since the demand for consumer services will fall if incomes earned in goods production fall. Further, economic growth and high living standards often depend heavily on exports, especially of goods. It is argued that services, by and large, cannot be expected fully to replace goods as the chief source of export income for all major industrial regions.

Among the questions posed by such considerations is, of course, whether or not old regions *are* de-industrialising, in the sense of experiencing a real loss of total goods production potential. The measures of loss commonly used include employment and shares of GNP. But both must be used with care, for they can conceal as much as they reveal.

Manufacturing employment has certainly declined relatively and, in recent times, absolutely in America, Britain and elsewhere among the old regions. This decline may well continue. Sectors of employment growth in manufacturing seem unlikely to offset fully the continued decline of jobs in older types of manufacturing (see Table 6.2 projections). But the employment record is only one part of the story – albeit a very important part, since, in the end, economic growth is desirable not simply for its own sake, but rather for its positive effects on people's lives. The falling employment in manufacturing, however, does not necessarily reflect how much is actually being produced, or how much wealth is being created. The benefits of increasing labour productivity must be kept in mind (as in the case of agriculture).

Similar comments apply to the measuring of de-industrialisation by the declining share of manufacturing in GNP. Figure 6.3 compares sectors in the US economy in 1960 and 1987. A considerable decline in the share of manufacturing and powerful growth in services is again the pattern. But total GNP grew tremendously over those years, and the value of the actual output of the manufacturing sector also grew. Using constant 1972 dollar values, manufacturing output grew by about 150 per cent between 1960 and 1987. Since manufacturing employment rose by only 14 per cent over that period, the record is one of considerable productivity growth rather than of industrial decline. This increase in productivity also throws another light on the falling share of GNP provided by manufacturing shown in Figure 6.3. Rising productivity means, in effect, lower production costs and, therefore, lower prices. The prices of manufactured goods, overall, have fallen considerably relative to services, so that their share of GNP was bound to fall.

One further point must be made. Types of employment within manufacturing industry itself began to change long ago. For some decades now production line (blue collar) workers have been declining in relative importance. Most growth has been in non-production (white collar) workers – supervisory and administrative personnel, sales and service employers, professional and technical staff, and so on. In 1947 such non-production employees accounted for 17 per cent of all manufacturing jobs in the USA; by 1986, for 36 per cent. Observation of this fact does not suggest a decline in American manufacturing. Rather it is a sign of increased sophistication and advancing technology. A far more 'flexible' production structure is being developed. But such changes have been affecting the comparability of employment data over time. Firms have been 'buying in' more of these white collar services from specialist outside service firms. This means that the jobs concerned, hitherto provided internally and thus counted as 'manufacturing' jobs, become bought in as 'services'. Thus some of the shift to service employment simply reflects a change in the operations of manufacturing firms.

In sum, old industrial regions like the USA possess complex economies within which a multitude of activities are interrelated and interdependent. Flexible methods of production make it impossible to tell precisely where goods production ends and service production begins. Thus judgements about a 'de-industrialisation' process are not easy to make. But what is very clear is that the service economy has not grown autonomously. More efficient goods production has played three essential roles:

1 Creating the higher incomes to promote demand for consumer services.
2 Creating an increasing demand for producer services.
3 Freeing labour to work in service occupations.

It follows that a true de-industrialisation experience, in the sense of a substantial loss of productive potential in manufacturing as a whole, would be a very serious matter. In the American case, however, there has been no such overall loss, although certain industries have declined. Today, fewer production workers can support many more in other jobs. The economic texture of society has 'thickened', with more layers of services, new job opportunities and higher incomes for most of the population.

But, as was stressed earlier, the transition experience can be painful. 'People problems' of adjustment to change are often associated with 'place problems' of large distressed areas. The United States has its share of these. So far, American experience has been examined as if it were a single 'old region' in the world economy. But the United States is not a single region; rather, it is a huge and varied country within which different parts have been affected differently by the changes outlined.

OLD AND NEW REGIONS OF THE USA

Within the United States we may justifiably regard the major industrial areas and communities of the north as 'old' regions, compared with the 'newer' regions of the south and west. During the nineteenth century, especially towards its end, the northern economy shifted emphasis quite rapidly from the primary to the secondary and its supporting tertiary sectors. Most manufacturing investment was concentrated in an area of the country bounded by Boston and Baltimore on the Atlantic coast and Milwaukee and St Louis in the interior.

This area became known as the 'manufacturing belt'. It remained dominant in the national economy until well after the Second World War. In 1950 the manufacturing belt identified by the American geographer, Edward Ullman, was calculated to account for 50 per cent of US population, but for no less than 73 per cent of all manufacturing jobs (Figure 6.4). Meanwhile the rest of the country (with important exceptions like southern California and parts of the Gulf coast) remained at an earlier stage of development, with large proportions of their working populations still engaged in primary activities. The more advanced economic structure of the north produced the higher incomes that sector theory suggests, so that it was by far the richest region of the country. In 1950 seven of the top ten states ranked by per capita income were still in the north while, by contrast, all ten of the bottom ranking states were in the south, where primary activities remained prominent. But subsequently the pattern began to change quite remarkably.

133

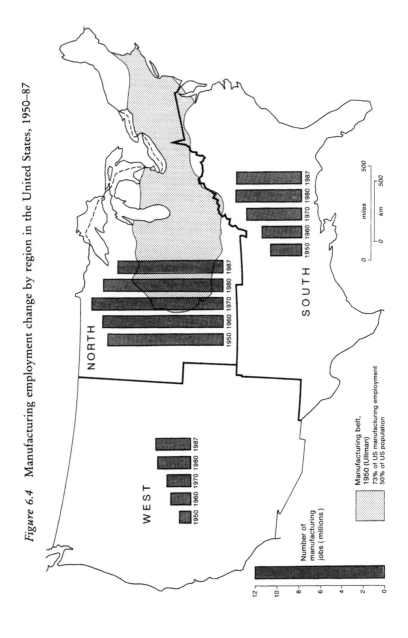

Figure 6.4 Manufacturing employment change by region in the United States, 1950–87

WEST

1950 1960 1970 1980 1987

NORTH

1950 1960 1970 1980 1987

SOUTH

1950 1960 1970 1980 1987

Number of
manufacturing
jobs (millions)

12
10
8
6
4
2
0

Manufacturing belt,
1950 (Ullman)
73% of US manufacturing employment
50% of US population

miles   500

0   km   500

Figure 6.4 shows the numbers employed in manufacturing in each of the three great regions of the USA (north, south and west) for 1950 to 1987. Table 6.4 gives the regional shares for the same dates. Total manufacturing employment grew in all regions up to 1970, but faster in the southern and western 'sunbelt' regions, so that the northern 'frostbelt' share fell. Then, between 1970 and 1980, industrial jobs began to decline absolutely in the north while growth continued in the 'newer' sunbelt locations. Boom conditions in the south and west contrasted with gloom and depression in many of the old industrial centres of the north. Textile and shoe manufacturing towns of New England, clothing centres of New York, iron and steel cities of Pennsylvania and Ohio, car-producing metropolises in Michigan and other mid-western states, all experienced plant closures, job reductions and general economic distress. The recession of the early 1980s brought manufacturing job losses to all regions (Figure 6.4), but the frostbelt was again hit more harshly, so that its share of jobs continued to fall.

Table 6.4 Regional shares of manufacturing employment, United States, 1950–87 (percentage of USA total)

Year	North	South	West	%	Total Millions
1950	72·4	19·5	8·1	100	14·8
1960	66·7	21·6	11·7	100	16·7
1970	62·3	25·7	12·0	100	20·4
1980	54·9	29·8	15·3	100	20·4
1987	51·4	31·4	17·2	100	19·1

Not surprisingly, these conditions were reflected in population change. The north experienced net out-migration of more than 8 million people between 1970 and 1987. These migrants did not simply follow the historical urge to 'go west'. The southern sunbelt became the chief beneficiary, being transformed from its historic role as the major centre of net out-migration into a region with massive net migration gains.

Explanations of these changes rest heavily on certain negative factors that are suggested to characterise old regions. These factors place them at a disadvantage compared with the positive attractions of newer developed, or developing, areas. Labour influences are often stressed, and are linked with the changing production requirements of many established manufacturing industries. Mass production

135

makes fewer demands on the labour skills that are present in old regions, while market pressures make labour cost savings essential. But wage rates tend to be high in old regions and total labour costs are raised further by liberal fringe benefits. Old customs and shopfloor practices may be hard to change, and labour is suggested to be resistant to innovation and more militant in its reactions to management pressures. A strong presence of trade unions is a common feature of old industrial regions. This is sometimes interpreted by investors as reflecting a poor 'business climate', whereas the relatively weak position of unions in most new regions is taken as a favourable sign.

Management, too, is often accused of ineptitude and inadequacy in old regions, especially in the old established firms. Lack of foresight, failures of judgement, complacency, slipshod methods, attachment to old ways, are all quoted, and contrasted with the vigour, the awareness of opportunities and the modern managerial outlook characteristic of new enterprises in new regions.

In addition to such human problems, the general environment of mature industrial centres may be less attractive. Old industrial buildings, hemmed in by other developments, cannot easily be adapted to new needs and purposes. Taxes may be high, but the public infrastructure may be old and dilapidated and public services of poor quality. Crime rates become high, journeys to work in large metropolises longer and more difficult, costs of living high. If all this be true (and, happily, it is not everywhere so!) the old region is in serious trouble.

But there is another side to this. It is important to recognise that the plight of old regions is also in large part due simply to the increased attractiveness of new locations – as in the US south (or overseas). Not only are labour costs lower and union activity much less, but there is a lower density of development, lower land costs, lower living costs, a warmer climate (made more attractive by air conditioning) the opportunities for a fresh start, and so on. Within the USA, too, the attractions of locations outside the north have been enhanced by a great improvement of national transport and communications networks, and by a pattern of federal government spending (especially for defence purposes) which has poured vast sums of money into the sunbelt areas of south and west. There was also, in the 1970s, the fortuitous impact of the energy crisis. This came at a critical time for the frostbelt – a major energy deficit area – and drew much investment capital and commercial attention to the oil, gas and coal rich regions of the south and west. All this was reflected in the turn-round in population migration mentioned above. Further, with the usual benefits accompanying its changing economic structure, the once backward south made great strides in closing the income gap that had separated it from the rest of the nation since the Civil War of 1861–65. In 1940 *per capita* income in the south was still only 66 per cent of the American average; but by 1987 it was 90 per cent. Of course, not everywhere in the south has been lifted to the same extent. Much poverty and backwardness remains, especially in rural areas. But rising average incomes have increased the region's attraction as a market location for many enterprises, thus helping to create a benevolent spiral of growth.

Given the catalogue of problems and the powerful competition of new areas of growth, the outlook for the old regions might appear dismal. But in fact they possess strengths of their own, some of which have already been mentioned. These can be of value in creating new roles and stimulating new phases of growth. So far, these strengths have not been much in evidence in all northern locations. But major changes have been taking place. One example of this is New England, which is discussed further in Chapter 7. Experience here in the 1970s and 1980s appeared to offer some grounds for optimism. The record of areas like New England explains more fully the nature of the problems of old regions and offers some signposts for other regions in the USA, and elsewhere, in the development of new roles.

QUESTIONS

1 What are the main stages of economic growth and which economic sectors are most important at each stage?

2 What is de-industrialisation and how can it be measured?

3 How has the 'production potential' of US industry changed since the 1960s?

4 What factors stimulated the shift of manufacturing between US 'frostbelt' and 'sunbelt' regions?

5 Describe the differences between the three 'great regions' (Figure 6.4) in the size and trends of manufacturing employment.

FURTHER READING

Chinitz, B. (1986) 'The regional transformation of the American economy', *Urban Studies* 23, pp. 377–85. Describes the dramatic changes in the fortunes of American regions in recent times, especially in the role of manufacturing industry.

Gillespie, A. E. and Green, A. E. (1987) 'The changing geography of producer services employment in Britain', *Regional Studies* 21(5). Examines the factors influencing the changing geography of producer services and the strength of the southern regions of the UK.

Green, A. E. and Howells, J. (1987) 'Spatial prospects for service growth in Britain', *Area* 19(2), pp. 111–22. Reviews the importance of the service sector for regional growth in Britain and examines the spatial prospects.

Keeble, D. (1987) 'Industrial change in the United Kingdom', Chapter 1 in Lever, F. (ed.) *Industrial Change in the United Kingdom* (Longman: Harlow). Examines recent sectoral and employment trends and their consequences for the industrial geography of the UK.

Watts, H. D. (1987) *Industrial Geography* (Longman: Harlow) Chapter 1, 'The geography of industrial change' (especially pp. 1–14). A wider view of industrial change and 'de-industrialisation', but with special attention paid to US and UK experience.

7

CASE STUDIES OF ECONOMIC CHANGE IN ADVANCED REGIONS

Robert Estall and Robert Bennett

STAGES, GROWTH AND CHANGE

Economic change, and the changes in productivity and economic structure that accompany it, are the essential ingredients of the prosperity of regions. As we have seen in the preceding chapters, economic change goes through cycles and stages in which different regions participate more or less successfully at different times. There are always new developments and changes taking place that continuously produce new regional patterns of development.

These pressures for change do not decrease with time. Indeed, they may increase. A cycle of development exists in many industries that can lead to the 'passing on' of old established industries to newer regions that offer superior conditions for production. At the same time 'new' regions emerge, based on innovations in products and technologies. Such changes now take place at a much faster pace than formerly. The 'product cycle' of manufactured goods from their development and first use to maturity and, perhaps, decline can sometimes be very short indeed. Some modern products in the field of electronics, for example, have been outmoded by new developments in research laboratories almost before they had established themselves on the market. Such a pace of change can be exciting; but it creates extra difficulties and uncertainties for both old and new regions.

In this chapter we examine some case studies of regional change. These case studies offer three contrasting experiences:

1 New England: the adaptation of an 'old' region.
2 Silicon landscapes: the development of 'new' regions, as in

California, eastern Massachusetts, Japan, south-east England and Scotland's 'Silicon Glen'.

3 Development from below: small enterprises and co-operatives, exemplified in 'Third Italy' and Spain's Mondragón co-operative.

NEW ENGLAND: ADAPTATION OF AN 'OLD' REGION

Industrialisation in New England began in a significant way some two centuries ago. With minor fluctuations, vigorous industrial growth then continued into the early twentieth century. In sector theory terms, the region was well advanced by 1900, when it had only 14 per cent of all its jobs in the primary sector (compared with 40 per cent in the USA as a whole) and 48 per cent in manufacturing (25 per cent in the USA). At that time it was among the richest of

Figure 7.1 Changing employment patterns, New England, 1947 and 1983

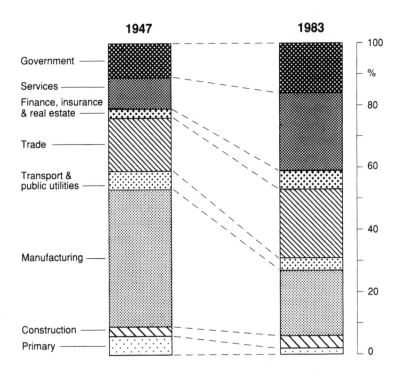

American regions, with a per capita income 35 per cent above the national average.

In this century, however, it faced increasing problems. As a highly specialised centre of manufacturing it was badly affected by the economic depression of the 1930s. Then (after a temporary boost created by the Second World War) it entered upon a period of difficulty and decline which lasted until the mid-1970s. Figure 7.1 depicts the changing employment patterns of New England in recent decades, and illustrates the dismal fortunes of its major manufacturing trades. The sharp relative decline of manufacturing employment illustrated in the diagram became an absolute decline after 1953, as Table 7.1 shows.

Table 7.1 Non-agricultural employment: New England and the rest of the USA, 1947–87 (000s)

	New England			Rest of USA[2]		
Year	Manufact.	Services	Total[1]	Manufact.	Services	Total[1]
1947	1,543	1,660	3,331	13,747	23,562	40,526
1953	1,600	1,838	3,584	15,659	27,075	46,076
1975	1,301	3,188	4,657	17,022	51,157	72,288
1987	1,369	4,705	6,395	17,696	76,821	95,915

Notes: [1] Including construction and mining.
[2] USA totals, minus New England.

The economic difficulties of New England were largely a result of labour, management and general 'business climate' conditions, which raised production costs. But the region was particularly hit by changes in the market for its major products of textiles and shoe and leather products. Woollen and cotton textiles were in a very real sense the 'engine' of early industrial growth in New England, and remained so into the present century. In 1919 the textile mills employed 442,000 workers, 30 per cent of all manufacturing employees in the region. Inter-war depression and the changing location of the industry in favour of southern states (especially the Carolinas), had reduced the number to 282,000 by 1947. Subsequently, new competition from the Third World and NICs reduced employment further – to only 50,000 workers in 1983, about 3 per cent of the region's manufacturing total.

For the great shoe and leather product industry the record was also disastrous. In 1919 156,000 were employed, some 10 per cent of all manufacturing workers. The subsequent decline of this industry

Plate 7.1 An old textile mill looking for a new role, Lowell, Mass., USA
(*Photo*: Robert Estall)

in the face of competition from other American regions, and imported footwear, reduced employment to 51,000 by 1983. These two great industries (textiles, shoes and leather) in their heyday had provided four out of every ten manufacturing jobs in the region. Today they are quite minor parts of New England's industrial structure. Small wonder that the intervening years saw much of the region in desperate economic circumstances.

But the region's misfortunes appeared to bottom out in about 1975. Till then, the continuous losses of the old staples more than offset employment growth in other sectors. But, from the mid-1970s, the growth of new types of manufacturing and a rapid expansion of specialised services allowed New England to take on many aspects of a 'new' region. This change has drawn strongly on the traditional skills and industries of the region, but adapted them to new needs. Among the traditional interests supporting this new phase of regional growth were the machine tool, metal product (including tools, cutlery and armaments) and instrument industries.

Plate 7.2 New factory age: a modern electronics plant, Andover, Mass., USA (*Photo*: Robert Estall)

These had provided a tradition of labour skills, innovatory habits and marketing techniques that were to be important for recovery. The infant electronic product industry also found a base in New England, where original scientific contributions in the electrical field were made early this century. Such interests provided a platform on which the region could build a new manufacturing structure based essentially on 'high-technology' activities.

As a consequence, the electrical and electronic equipment industry has become the largest manufacturing sector in New England (as it is in the USA), with a specialised focus in communications equipment and other electronic devices and accessories. Outstanding, too, has been the growth of computer and related equipment manufacture. Some 14 per cent of all US employment in computer manufacturing has been calculated to be in New England – twice the region's share of total manufacturing employment.

Other success stories include the transport equipment industry, now largely concerned with the manufacture of aircraft engines and parts, helicopters and nuclear submarines. Instrument manufacture is another major field of high-technology endeavour, and scientific instrument production in New England probably accounts for some 15 per cent of national employment in this industry. In sum, the pattern of New England's manufacturing activities has been transformed, and about half of present New England manufacturing jobs are found in the 'high-technology' sectors. In addition, the erosion of manufacturing jobs came to an end, and the years from 1975 to 1987 saw modest growth (Table 7.1).

Let us examine the conditions affecting this turn-round. The catalogue of disadvantages increasingly affecting the New England economy this century indeed makes formidable reading. Remoteness from growing internal markets; lack of industrial raw materials and, especially, a high cost of energy; old plant and equipment and complacent management in many long established firms; ageing community infrastructure and a depressing environment in old industrial cities; a highly unionised and militant workforce; high wages in major old staples seeking to lower their labour costs; high cost of living; a reputation for high taxation. Such disadvantages (never equally true of all parts of the region, of course) understandably affected the decisons of investors.

But such conditions are not necessarily unchangeable. Some have been much modified in recent decades, with, for example, sharp falls in levels of unionisation, reform of taxation and redevelopment of city areas. It is also the case, stressed in Chapter 6, that old regions possess positive characteristics that can eventually triumph over the negative ones. In New England these include a skilled and experienced workforce (for which wage levels often became lower than in competing American locations). There is also the major research and development capacity that has clustered around the region's prestigious universities and other centres of advanced education. Outstanding, of course, are the Massachusetts Institute of Technology (MIT) and Harvard University in Boston. But these are only the highlights of a truly major concentration of advanced scientific and technological education and training provision in the region. This provision has been associated with a new vigour in innovation, which is itself often supported directly by the universities and research institutes and by the ready availability of capital for new ventures.

These changes have been accompanied by vigorous tertiary sector growth. Figure 7.1 shows the region's employment structure in the 1980s to be dominated by services. These accounted for 74 per cent of all jobs in 1987. Notable expansion has come in such services as banking and insurance (where, again, New England has a long established reputation) and in various kinds of business and professional services. Hospital and health services in the region are also highly regarded nationally, so that this sector has had the fastest rate of growth of all, with employment in it now approaching half a million persons. Also, about one in every eight persons engaged in private education services in America lives in New England.

Direct 'exports' of many services also help to sustain the region's income and to replace the lost manufacturing staples. Tourism, for example, is now a major industry. The region possesses many physical attractions in its glaciated mountain landscape, its many lakes and its glorious coastline. It also retains much 'colonial' charm in its small townships. Employment created in this activity is scattered widely through all sectors of the tertiary group. New England insurance, banking, educational and medical services also figure very prominently among its modern earners of export income.

The crucial role of the Second World War, the Korean War and the Vietnam War in this 'restructuring' should perhaps be stressed. The machinery, electrical (electronic) equipment, aircraft engine, helicopter, instrument and other technologically advanced industries have been greatly stimulated by government defence demands; which also powerfully reinforced the region's capacities in industrial research and development. But this considerable government spending role has caused reservations to be voiced about the causes and the character of recent growth. New England's industrial recovery since the mid-1970s has rested heavily on a massive increase in federal government defence spending. This brought large new contracts for research and development of sophisticated weapon and defensive systems in the so-called 'Star Wars' programme. Such spending will not be maintained at the 1980s levels, given the changed international climate. There is also uncertainty about the long term contribution of high-technology industry to job creation. New products and processes, once established in the market place, can be siphoned off to new locations, while the continuing application of new technologies to improve productivity in established industries will probably continue to reduce direct employment in manufacturing. Total manufacturing employment in

New England, despite recent recovery, remains well below earlier levels (Table 7.1), and the longer term outlook, as for the nation as a whole, is probably slow decline.

Again, it can be argued that the increased dependence on service jobs has not made the region less vulnerable to economic change. Much service employment depends directly on manufacturing, while current exports of services are vulnerable to the development by other regions of their own specialist service trades. Reservations are also expressed about the recent employment record. It is suggested that the very low rates of unemployment in the region in the late 1980s were partly due to low rates of population (and thus labour force) growth. The pressure on labour supplies that this causes will inevitably be reflected in rising labour costs and declining competitiveness in future years.

Thus views on the New England revival are by no means unanimously optimistic and the region enters the 1990s in a condition of some uncertainty. None the less, it stands as a major example of an old economy discovering new strengths to carry it into a new stage of growth. Indeed, the experience of New England can serve as a model of the particular strengths of an old region. These include a labour force of good quality; an educational system that is trying to support and reinforce that quality; an abundance of scientific and technological expertise; an established research infrastructure of a 'critical mass' that permits the interchange of ideas and provides necessary technical support for development and innovation; investment capital and supporting financial institutions; a wealth of business and entrepreneurial experience. From such a list it becomes clear that the basic resources of an old region are related to the skills and energy of its people.

SILICON LANDSCAPES

This term has been applied by the geographer Peter Hall to the regional changes which were first experienced in California's Silicon Valley and around Boston's Route 128 freeway in Massachusetts. Some of these changes have underpinned New England's transformation, outlined above. Similar developments have subsequently been identified in other countries: in Japan, in central Scotland ('Silicon Glen') and in the Maidenhead–Newbury area of southern England ('Software Valley').

The concept of 'silicon landscapes' has been derived from the emergence of the silicon chip and the microprocessor as key elements in a modern economic landscape. Hence it has depended specifically on the innovation and new product development arising from technological advance. The impact of these changes has been focused on high-technology industries which have required high innovation inputs. But Hall and others have also stressed the crucial significance of government support. For example, in the USA and Britain there has been a major role played by defence contracts and research commissioned for the defence industries. We have already noted this in the case of New England. Federal government contracts have also stimulated a very high rate of technological advance in other centres, and perhaps most notably in southern California.

In the USA, state and local governments, too, have often attempted to 'oil the wheels' of innovation, with grants and local tax concessions. Each has sought what Hall calls 'a new economic Holy Grail – industrial renaissance through high-technology job creation' where every city and region sets up its technology park or university 'science park'. For example, the US business magazine *Venture* in 1983 identified fifty regional high-tech complexes in the USA. Of these California's Silicon Valley, Boston's Route 128 and North Carolina's Research Triangle Park were considered 'mature', whilst thirty-two other high-tech complexes were classified as 'developing'. Of these the most promising were Louisiana's Silicon Bayou, Maryland's Satellite Valley, Virginia's Dulles Corridor and Florida's Silicon Beach. The remaining high-tech centres were 'emerging'.

Hall maintains that the success of these complexes depends on the capacity to innovate and 'stay one step ahead of the action'. He argues that regions that are successful in doing this create a general intellectual environment of R&D external to each firm, which in turn creates the momentum for further innovation and research. As a result, the agglomeration of new R&D facilities creates an upward spiral of employment and economic growth in the successful centres. Many of the most successful innovations are made by new, small firms which are closely linked to this wider technological environment. In California's original Silicon Valley most of the firms have fewer than 1,000 workers (see Figure 7.2). But their earnings can be very large, despite their small size. Hall notes that the revenue from computer software products in the USA grew by 42 per cent between 1980 and 1985, and for computer professional

Figure 7.2 Firms in Silicon Valley, California

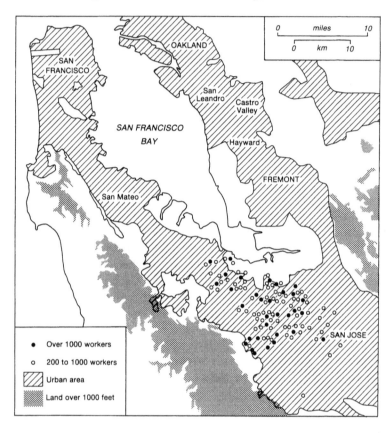

services by 27 per cent. In biotechnology even more spectacular growth is taking place: for example, employment in California in this sector grew from only 279 in 1978 to 3,064 by 1982.

The continued growth of these high-tech areas depends upon their continuing capacity to innovate. In this process, some regions appear to have developed a superior capacity to others. For example, survey evidence for the period 1976 to 1982 shows that Silicon Valley in California maintained a rather higher rate of innovation of new products than did Britain's two main high-tech areas (in Scotland and south-east England). In Silicon Valley 85 per cent of firms had innovated new products in this period, whilst in

148

south-east England the figure was 78 per cent and in Scotland 63 per cent. In addition, Silicon Valley firms appear to have a greater impact on other local firms, thus increasing the agglomeration effects and cumulative advantage. In Silicon Valley 68 per cent of firms purchased over half their material inputs locally. Only 18 per cent of the south-east England firms achieved this level and a mere 11 per cent in Scotland (see Hall, Breheny et al., 1987).

Despite the slower rate of innovation and much lower local multipliers in the two British cases, there is evidence of rapid growth in these areas. They may even be catching up on the Californian leader. Figure 7.3 shows the rapid job growth in south-east England, particularly in the Thames valley. On the other hand, there have also been jobs losses in central London and the older industrial areas as part of a notable decentralisation trend from large cities as a whole. Hence, whilst innovation centres are offering a glimpse of the new landscape of future geographies, they are also tending to reinforce forces that concentrate growth in certain 'preferred' locations. The consequence for the immediate future is, therefore, likely to be continued imbalances between regions as these new regions grow and older regions struggle to adapt.

Because of the difficulties of adaptation and the very high costs of advanced R&D, governments have had to become increasingly involved in high-technology policy. This involvement has been greatly stimulated by the needs of national defence industries. This is certainly true of California, New England and south-east England. However, governments have also been involved because they are often the only sources of the long term research funds necessary to finance the enormous R&D costs of new technologies. Governments also seek to serve the national interest and help their countries to cope in the global game of economic competition.

A major example of government intervention on a much larger scale to stimulate change and help the national industrial base is provided by *technopoles*. This term has been applied to attempts to create growth poles in specific regions or localities by combining high-technology innovation centres with local or regional governmental support. Germany and France provide leading examples in Europe. Britain and other European countries have also registered some success. But Japan offers the main example of this kind of government-led innovation centre. The key to a technopole is the rapid development of R&D around a major research institute or

Figure 7.3   Numbers employed in high-technology industry, Great Britain, 1975 and 1981

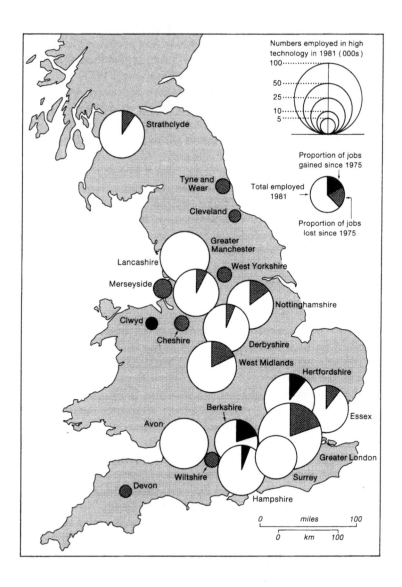

university, and the supportive environment of a highly skilled local labour supply. Local, as well as central, government support is crucial to these areas, since few private sector bodies can afford to fund long-term research costs and product development for commercial exploitation. Many projects also need the support of experts employed in government research centres, particularly to use the spin-offs from defence industry research.

In Japan, central government ministries and local government initiatives have both been key factors in stimulating development, first in Kyushu Island, then in Kumamoto and now in many other areas. The prefecture government of Kyushu undertook a number of initiatives to attract electronics firms in the 1970s. A national programme of government research stimulus also helped these industries. Today nearly 40 per cent of all Japanese production of integrated circuits comes from Kyushu.

The keys to the success of this Japanese approach have been:

1 The use of *branch plants* to stimulate local small and medium-sized firms.
2 A high quality *local labour supply* with the right skills base.
3 Generous *government funding* of local research centres.
4 Major further *financial support* (subsidies, incentives, infrastructure provision, purchase of equipment) from local and central government.
5 Establishment of *applied industrial research institutes* for the exchange of ideas, and the development of common research and consultancy advice.

Since 1983 twenty-five such technopoles have been approved by the Japanese central government throughout Japan (see Figure 7.4). These represent a national system of decentralised high-technology growth centres. The performance of other countries does not match this success, except for a few examples in the USA (e.g. around the Massachusetts Institute of Technology in Boston) and in Korea. On the other hand, a number of German and French universities and Chambers of Commerce have been successful in stimulating 'science park' developments on a smaller scale, to the benefit of the regions in which they are located.

Figure 7.4 The location of technopoles, Japan

Source: Adapted from I. Masser (1990) 'Technology and Regional Development Policy: a review of Japan's Technopolis Programme', *Regional Studies* 24, p. 45

DEVELOPMENT FROM BELOW: SMALL ENTERPRISES AND CO-OPERATIVES

The changes we have discussed in the previous two case studies have been to a large extent 'top down' experiences of regional change, where national or global markets and government activities have stimulated a response from local levels. But regional change can also be stimulated with a greater emphasis from below. In this case networks of very small firms can become interlinked in a complex web of relationships to provide a major engine of growth, using the innovative capacity and diversity of a small firms base. Two contrasted examples of this experience are outlined here: Third Italy and the Mondragón co-operatives.

Small enterprises

'Third Italy' is a term used to describe the recent experience of economic growth in an area of north central Italy around Bologna, Florence, Ancona and Venice. This lies outside the main traditional Italian industrial centre of 'first' and 'second' Italy, Milan–Turin. Within this area a network of small enterprises, often with ten or fewer workers, is spread through many villages and small towns. The firms are often technologically advanced and cover a wide range of products from consumer goods, such as shoes, motor cycles, clothing and textiles, to machine tools and component manufacture for larger industrial concerns.

The key aspect of Third Italy is its organisational structure. The chief characteristic is informal, but close, co-operation between individual enterprises. This is shown schematically in Figure 7.5 as a network of linkages between designers, owners, technicians and workers for individual firms. Such people will not normally work within one building, but will be scattered among many locations, including people's homes. There are also close links between firms in pursuing the R&D for new products, in filling orders, in marketing, purchasing materials, in obtaining financial capital, and in technical and consultancy services. The result is a unique mixture of styles which resembles both 'co-operatives' and 'cottage industry'.

Third Italy has other special characteristics, of which two are particularly important: the high level of innovation and the presence of facilitating conditions. The *high rate of innovation* is difficult to explain, but appears to be the result of the capacity of very small firms to respond to change. In responding, however, they quickly reach constraints of size and require help from other linked firms who, in turn, require support from their linked firms. This has stimulated a high level of mutual collaboration and partnership between firms, illustrated by the intensity of contracting and subcontracting.

The *facilitating conditions* appear to be the result of four further factors. First, a relaxed regime of tax administration and low enforcement of employment regulations. This has permitted

Figure 7.5 The organisational structure of Third Italy

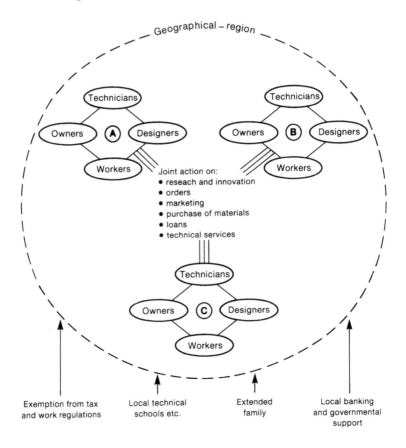

individuals to achieve higher incomes by avoiding tax, and to use labour inputs (including family relatives) in a far more flexible pattern than would otherwise have been possible. Hence the 'black' or 'shadow' economy has been particularly significant. A second factor has been the Italian cultural tradition of the extended family. This has provided a basis of trust, contacts and support between linked enterprises that would not have occurred so easily by other means. Third, local technical schools and training establishments have seized on the possibilities of this new labour market by rapidly initiating new courses and, indeed, influencing some of the innovative developments. A fourth factor has been the support of local government, local banks and other institutions in providing the necessary credit and financial facilities.

Co-operatives are another approach to such 'development from below'. Co-operatives are not a new mode of development but have existed, particularly for agricultural producers, for many years. Co-operatives may, however, offer one of the best means of 'development from below' in the Third World, some parts of NICs and in the rural areas of advanced economies such as southern Europe. They combine self-help solutions with considerable incentives to individuals and with the scope to organise individual action into viable size units to attain economies of scale. At larger scales it is also possible for government or venture capital funds to reach co-operatives that need support, whereas for large numbers of very small producers it is often difficult to obtain financial capital.

The co-operative approach not only offers a means forward in agricultural development. It has also been used successfully in manufacturing and in high-technology industries. In this context it provides opportunities not only for Third World and newly industrialising countries, but also for the industrialised world.

Perhaps the best, and most famous, example is the Mondragón Co-operative. It was founded in the 1940s in the Basque country in north-east Spain and has become the leading example of worker ownership in the world. In the 1980s it embraced 160 co-operative enterprises, with 19,000 members. These represent 2·5 per cent of the active working population of the area, which is spread over a wide geographical area of small and medium-sized towns in north-east Spain and south-west France (see Table 7.2). A variety of agricultural and manufacturing sectors are represented, ranging from metal working to consumer goods, and from industrial services, housing, community services, education and training, to retail outlets. The co-operatives have also developed their own group bank, the Caja Laboral Popular. This has been so successful that it now attracts funds in excess of the needs of the co-operatives.

The classic studies of this co-operative by Wiener and Oakeshott and Thomas and Logan argue that the worker–management co-operation which has developed has allowed innovation and development within a set of dispersed but co-ordinated firms. This is similar in many ways to the Third Italy, but with somewhat larger units, and with a much wider set of social objectives, especially relating to organised training and research. Like Third Italy and the other

Table 7.2 Development of the Mondragón experience, 1960–85

Year	No. of members	No. of co-operatives	Sales[1]	New investment[1]	Deposits of Caja Labora Popular
1960	395	4	–	–	0·1
1965	3,441	36	17·5	–	3·4
1970	8,543	52	50·8	7·1	23·1
1975	13,808	65	80·1	11·7	59·8
1980	18,733	92	119·5	8·2	88·9
1985	19,200	111	141·0	7·5	147·0

Note: [1] Thousand million pesetas at 1985 prices.
Source: H. Wiener and R. Oakeshott, *Workers–Owners: Mondragón Revisited*, Anglo-German Foundation: London, 1987

Figure 7.6 Co-operative interrelationships, Mondragón

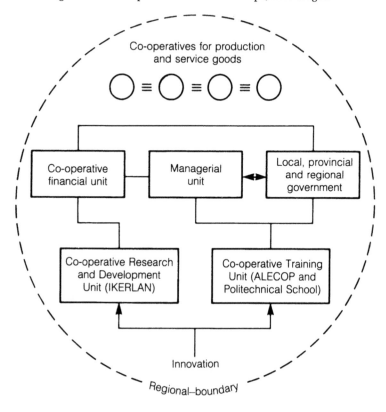

regional developments we have discussed, the role of local government and other local actors, such as the universities, has been crucial. Individuals are also sent for training to foreign universities and research centres.

These interrelationships are shown in Figure 7.6. In recent years, although not in its original foundation, modern technologies have been a key part of the success of the co-operative, allowing it to develop and grow rapidly (as shown in Table 7.2) whilst the traditional industrial sectors in the Basque country have declined. Key sectors involved in recent growth have been computer-aided design, robots, and electronics. But there has also been continued growth in consumer electrical goods and other sectors.

The special cultural and ethnic character of the Basque country, now recognised under the new Spanish constitution as a special regional government, has ensured a special regional growth impetus for the co-operative. The earnings of the co-operative are almost all reinvested locally. Thus for the continued survival and prosperity of the co-operative a continued process of innovation is required, together with tough economic decisions: only to support those investments which have most potential in international competition. Although local political factors have not always allowed these tough decisions to be taken, the co-operative has been remarkably successful in 'locking in' the earnings of its highly profitable industries to reinvestment for the continued economic expansion of its region.

CONCLUSION

The developing global interrelationship between places is creating the need for new roles to be defined for nations, regions, cities and localities. Both capital and labour markets are responding to these needs for change, and new geographies are developing by which these factors of production are assembled. Each region and locality has to work towards adapting itself to achieve the right circumstances for its economic success: in labour skills, finance and capital resources, in managerial capacity, and in the quality of its local environment. We have looked here at some case studies of regional change based on high-technology and organisational change. New England, silicon landscapes and development from below offer overlapping, as well as different, views of regional possibilities for adaptation. These offer examples which other areas might seek to emulate.

The examples we have discussed are not exclusive. If we are indeed experiencing a world scale revolution in economic relationships it is likely that a wide variety of new economic, social and political forms and practices will develop. The forms that these may take are not fully foreseeable but, whatever the pattern, successful national and regional strategies must aim at achieving the maximum capacity to adapt to future change. Once again, therefore, it comes down, in the end, to people. All will depend in the final analysis on how individuals and societies learn to cope with change.

QUESTIONS

1 What factors led to New England's economic decline and subsequent recovery?
2 Contrast California, Massachusetts (using the New England case study) and south-east England in terms of (a) factors stimulating development, (b) types of products, (c) types of suppliers, (d) types of customers for products.
3 Select the key factors that have helped the success of Japanese technopoles and Third Italy.
4 What are the key elements in the success of the Mondragón cooperative?

FURTHER READING

Dicken, P. and Lloyd, P. E. (1981) *Modern Western Society: A Geographical Perspective on Work, Home and Well-being* (Harper & Row: London). A useful review of the factors affecting growing and declining regions.

The Economist, 8 August 1987, 'A survey of New England'. An accessible and well presented discussion of new developments and problems overcome.

Hall, P. and Markusen, A. (eds.) (1985) *Silicon Landscapes* (Allen & Unwin: London). Excellent account of changing economic patterns in relation to product cycles, together with studies of Silicon Valley and the UK M4 corridor.

Hall, P., Breheny, M., McQuaid, R. and Hart, D. (1987) *Western Sunrise: The Genesis and Growth of Britain's Major High-tech Corridor* (Allen & Unwin: London).

Paterson, J. H. (1988) *North America*, 8th edn (Oxford University Press: Oxford) Chapter 11, 'New England'. A very readable account of the geography of this important region.

8

DEVELOPMENT AND THE THIRD WORLD

Sylvia Chant

DEFINITIONS

The term 'Third World' is normally used of those regions which are less developed than advanced capitalist countries, such as the UK and the USA (the 'First World') or what have until recently been centrally planned economies, such as the USSR or Eastern Europe (the 'Second World'). Geographically speaking, the Third World covers Latin America, the Caribbean, Africa (excluding South Africa), the Middle East, South and South East Asia, and most of Oceania except Australia and New Zealand. Given that all these regions lie south of northern America, Europe and the USSR, the term 'South' is also used to refer to the Third World, especially since the publication of the Brandt Report (1980) *North-South: A Programme for Survival* (see Figure 4.2).

The term '*less* developed' itself is, of course, relative. It also embodies a varied range of economic, demographic, social and political characteristics which may not always apply, or apply in equal measure, to all Third World countries. The conditions used to describe such countries normally include, on the one hand, low levels of industrialisation, urbanisation, per capita gross national product, life expectancy, literacy and nutritional intake and, on the other hand, high rates of infant mortality (alongside population growth), poverty, inflation, reliance on primary export products, and per capita debt. Another notable feature of Third World countries is that most have been subject to some form of colonial rule in the last 500 years.

A commonly used development indicator is the level of per capita GNP. This gives a measure of income disparities between countries and was discussed in more detail in Chapter 4. Richer countries such as the USA and Japan boast per capita GNPs (1988 levels) of above US $19,000 and US $21,000 respectively. Some of the poorest, such as Haiti and Bangladesh, have only US $380 and US $170 respectively.

While per capita GNP has long been used as the main tool for measuring development, in 1990 the United Nations proposed its replacement by a composite measure of economic and social well-being known as the *human development index* (HDI). This takes into account average per capita purchasing power within each country, together with levels of literacy and life expectancy. Use of the HDI is argued to provide a more rounded measure of development than one which concentrates solely on economic performance. Conversion of a country's gross national product into hard currency (US dollars) may mean little in terms of what people can actually afford to buy, while GNP measurement may fluctuate dramatically in the light of variable exchange rates. Figure 8.1 shows four major HDI groupings for the 130 countries in the world with a population of 1 million or more. The HDI tends to raise the development status of countries which may be rather poor, such as Jamaica and Costa Rica, but which have committed resources to public expenditure on health and education. It also draws attention to the lower human development achievements of some richer Third World countries such as the oil-exporting states of the Middle East, where wealth remains concentrated in the hands of a few. Overall, however, it is apparent that in general, high per capita GNP is usually accompanied by higher levels of social development, so that the map of human development contains few surprises.

In free market economies, high per capita income is also usually linked with high levels of *industrialisation*, itself an important contributor to economic growth, as shown in Chapters 5 and 6. In France and the Netherlands, for example, 37 per cent of gross domestic product is provided by manufacturing industry, compared with only 18 per cent in Malawi, 14 per cent in Bangladesh and 9 per cent in Somalia. (GDP is the total earnings or product of the economy excluding overseas earnings.)

Urbanisation is another commonly used indicator of development, since, on the whole, it displays a positive relationship with levels of industrialisation and per capita income. One estimate suggests that there is a correlation coefficient of 0·75 between GNP per capita and the percentage of the population in urban areas. But definitions of what constitutes an 'urban settlement' vary

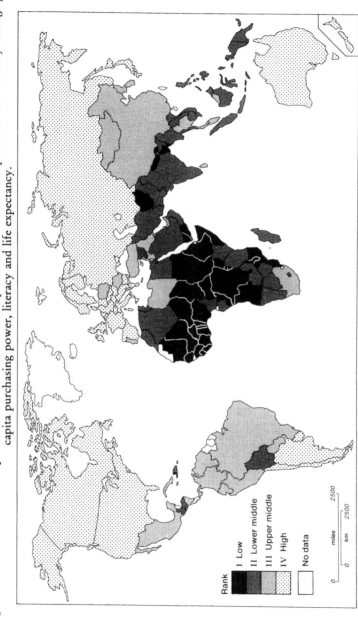

Figure 8.1 The human development index. The rank order depends on levels of development as measured by average per capita purchasing power, literacy and life expectancy.

Source: United Nations Development Programme, *Human Development Report,* 1990

Rank

I Low
II Lower middle
III Upper middle
IV High

No data

miles 2500
0

km 2500
0

considerably. In countries such as Liberia and Honduras any settlement of 2,000 inhabitants or more is classified as urban. In India, Ghana, Chad, Mali and Botswana the urban cut-off point is 5,000. More generally a threshold of 20,000 is used. But when one is concerned with establishing the proportion of people living in cities (as opposed to towns), international bodies such as the United Nations use a figure of 100,000. Differing definitions make it hard to compare levels of urbanisation between countries and continents but, bearing in mind this problem of comparability, the average level of urban population of all World Bank reporting members in 1988 was 47 per cent; with 78 per cent in high income economies and 41 per cent in low and middle income economies. This testifies to the fact that less developed countries are generally less urbanised.

There are obvious problems not only with most indicators of development but also with the use of the term 'Third World'. The most important include, first, the often implicit assumption that 'West is best', and that Third World countries should try to emulate the example of the industrialised economies. The term 'Third World' itself implies an inferior position in the global hierarchy.

Second, the terms 'First', 'Second' and 'Third' Worlds tend to foster an idea that they are separate entities, rather than parts of the same interlocking global system. Many would argue that the Third World exists only *because* of its dependent relationship with the advanced economies (see below).

A third problem is that, as some of the above figures have indicated, there are huge disparities between various regions and countries of the Third World which make the term rather too broad to be very useful. One major differentiating factor is politics. Socialist Third World countries such as Cuba and Vietnam may be fairly poor in economic terms but tend to have higher levels of social facilities, such as medical and educational provision, than many other developing countries. Another major differentiating factor between Third World regions is economic performance. Latin America, South East Asia and the oil-exporting countries of the Middle East are economically much more developed than, say, sub-Saharan Africa or various Southern Asian countries such as Bangladesh and Nepal. The former have higher per capita incomes, and higher levels of industrialisation and urbanisation.

Within separate Third World regions there are again vast disparities. For example, Latin American countries such as Brazil, Mexico and Argentina have comparatively high levels of per capita

GNP and a large percentage of the labour force in industry compared with Bolivia and Honduras, which are far poorer and where most of the economically active population is in agriculture (see Table 8.1). The same diversity is apparent in Asia, where Taiwan, South Korea, Hong Kong and Singapore have been particularly successful in achieving high levels of industrialisation and economic growth. In 1988 the latter two countries had per capita GNPs of over US\$9,000, contrasting markedly with other predominantly agricultural countries in the region such as the Philippines and Indonesia where per capita GNP was well under US\$1,000 (Table 8.1).

Within each Third World country, too, there are wide social differences. Urbanisation is a useful measure of this situation. Rural or urban residence plays a major role in influencing levels of economic and social well-being among the population. Contrasts between city and countryside are usually extremely marked, not only in visual terms, but also in terms of access to material and social resources. The provision of medical care, schooling and physical infrastructure such as roads and services is usually far better in cities. In Colombia, for example, the ratio of doctors per head of population is 1 : 1,000 in towns, but 1 : 6,400 in rural areas. This kind of phenomenon plays a major part in influencing national spatial variations in well-being.

APPROACHES TO DEVELOPMENT

With these various indicators of development in mind, it is now necessary to examine some of the main ideas put forward to explain how development, or underdevelopment, occurs, and what policy measures for improving Third World prospects might be most appropriate. There are several theoretical strands in the literature, although two major schools of thought may be identified: 'modernisation theory' and 'dependency theory'. These are discussed below, along with two more recent approaches known as 'alternative development' and 'basic needs'.

Modernisation theory

Modernisation theory emerged in the 1950s and 1960s in a surge of post-war optimism about the prospects for world peace and development. It emphasises the necessity of shifting from traditional to modern values in order to accelerate the development process.

Table 8.1 Development indicators for selected countries in Latin America and South East Asia

Country	Population (millions mid-1988)	Area (000 km²)	GNP per capita 1988 (US$)	Average annual growth rate of GNP per capita 1965–88 (%)	GDP 1988 (US$ million)	Distribution of GDP (%) Agriculture	Industry	Services	Percentage of labour force in agriculture 1980	Percentage of labour force in industry 1980	Life expectancy at birth (years) 1988
Latin America											
Argentina	31·5	2,767	2,520	0·0	79,440	13	44	44	13	34	71
Brazil	144·4	8,512	2,160	3·6	323,610	9	43	49	31	27	65
Mexico	83·7	1,958	1,760	2·3	176,700	9	35	56	37	29	69
Ecuador	10·1	284	1,120	3·1	10,320	15	36	49	39	20	66
Peru	20·7	1,285	1,300	0·1	25,670	12	36	51	40	18	62
Bolivia	6·9	1,099	570	0·6	4,310	24	27	49	46	20	53
Honduras	4·8	112	860	0·6	3,860	25	21	54	61	16	64
Cuba	10·0	111	–	–	–	–	–	–	24	29	76
South East Asia											
South Korea	42·0	98	3,600	6·8	171,310	11	43	46	36	27	70
Hong Kong	5·7	1	9,220	6·3	44,830	0	29	70	2	51	77
Singapore	2·6	1	9,070	7·2	23,880	0	38	61	2	38	74
Thailand	54·5	513	1,000	4·0	57,950	17	35	48	70	10	65
Philippines	59·9	300	630	1·6	39,210	23	34	44	52	16	64
Indonesia	174·8	1,905	440	4·3	83,220	24	36	40	57	13	61
North Korea	21·0	121	–	–	–	–	–	–	76	7	70
Vietnam	64·2	330	–	–	–	–	–	–	68	12	66

Note: Taiwan does not appear in the table because it is not listed separately by the World Bank. A dash indicates that no data are available.
Source: World Bank, *World Development Report 1990*, Oxford University Press: Oxford, 1990

As shown in Table 8.2, 'traditional' values are seen to belong to societies where people have little interest in the future, are fatalistic, superstitious and resistant to change, are oriented towards their families and local communities rather than the nation as a whole, and whose status is based on ascription rather than achievement, i.e. family of birth and position in the kin group are more important than individual attainment. In 'modern' societies, on the other hand, people are regarded as being open to change, less attached to their families, rational and entrepreneurial, and able to determine their position in society by their own work and achievements.

Table 8.2 Characteristics of 'traditional' and 'modern' societies

Context of action	'Traditional' response	'Modern' response
Orientation	Family/local community	State/nation
Social/political order	Custom/conformity	Innovation
Authority	Divine/sacred	Secular
Determinant of social roles	Ascription	Achievement
Nature of official relationships	Personal	Impersonal

Source: After C. S. Whitaker, *The Politics of Tradition: Continuity and Change in Northern Nigeria, 1946–66*, Princeton University Press: Princeton, 1970

Third World countries in the post-war period were seen to fit the bill of 'traditional' societies and, as such, were likened to the pre-industrial peasant communities of north-west Europe. In order to ensure a successful passage from tradition to modernity, it was thought, Third World nations would have to experience processes of social change similar to those of the West a century or two earlier. The idea that all countries would have the same experience of development and were merely on different rungs of the same ladder to economic prosperity was incorporated in 1960 into the stage model of economic development by W. W. Rostow. Stage one was 'traditional societies' which could, at some given moment, move into stage two, referred to as the 'pre-conditions for take-off' (comprising an increase in internal security, entrepreneurial activity and capital investment); stage three was 'take-off' itself, stage four was the 'drive to maturity' and stage five brought the 'age of high mass consumption'.

In setting out these stages of development, it was implicitly assumed that the historical experience of the West could provide a model for Third World countries. However, one key difference between seventeenth and eighteenth century Europe and post-war developing societies was the belief that the latter could be helped on their way to developmental success by the benign intervention of the advanced countries. This idea had far-reaching implications for the development policies adopted in the Third World from the 1950s onwards, and in particular between 1960 and 1970, the period of the United Nations First Development Decade, when increased levels of GNP became the primary aim of developing economies.

One major policy which derived from this commitment to economic growth was to promote more rapid change in Third World countries. This could be achieved by exposing them to the technology, practices and ideas of the West, and by diffusing development as far and wide as possible through trade and foreign investment.

A second implication was that Third World countries should imitate the Western development process and, in particular, that they should industrialise. Given that industrial development in nineteenth century Europe had been associated with a massive increase in urbanisation, in the Third World, too, it was imagined that industrial investment could be most rapidly stimulated in cities. Cities were not only an efficient way of centralising productive activities, but they could also help to break down the traditional social customs that inhibited economic progress.

Both rural–urban and interregional inequalities became necessary evils in this particular strategy of economic development. As a means of getting growth under way, money was poured into cities at the expense of the countryside, and into industrial regions at the expense of the non-industrial. However, according to modernisation theory, inequality would be only a short term phenomenon. In the future the benefits of growth in favoured areas would 'trickle down' into the 'backward' rural regions. Although many writers were sceptical that growth would automatically spread to all regions over time, these ideas remained pervasive. They ultimately became embodied in John Friedmann's 'centre–periphery' model, whereby cities were seen as mechanisms to stimulate economic change, using them as 'growth poles' that would gradually throw development impulses outwards to the periphery.

Criticisms have been levelled at modernisation theory's interpretation of development on a number of counts. First the use of terms such as 'traditional' and 'modern' have come under attack, this dichotomy being viewed as being present only in the minds of Western academics and planners and thus inapplicable to the real world. Moreover, many so-called 'traditional' values not only persist in 'modern' societies but actually operate to sustain development. One might cite here the importance of kinship networks in bolstering the survival of the urban working classes through providing shelter, job contacts and economic support.

A second major criticism of modernisation theory is that it was overly descriptive and never really stated convincingly *how* social change was to occur. Third, modernisation theory dwelt too much (and too naively) on the role of attitudes in influencing development. It tended to ignore the fact that attitudes themselves are shaped by external influences, which in the case of many Third World countries have involved negative experiences of colonial rule and severe restrictions in determining their own development goals and strategies.

Finally, modernisation theory has so far proved incorrect in its predictions of evolutionary economic progress. Although GNP growth rates showed dramatic rises in the 1950s and 1960s, subsequent experience has shown that this period of development was a purely economic phenomenon and tended to lead to considerable, and persistent, inequities. As a result not only have the vast majority of Third World countries remained peripheral to the world economy, but only a few people within those countries have actually benefited in any direct way from growth. Indeed, conditions for many of the rural and urban poor have actually worsened over time. These problems were recognised at the end of the 1970s by the Independent Commission on International Development Issues chaired by the West German Chancellor Willy Brandt, who recommended that, in order to alleviate world poverty, the wealthy North would have to double its aid expenditure and concede improved terms of trade to the South. The medicine prescribed by the Brandt Commission was, therefore, more North–South cooperation, not less; a treatment which, in the opinion of the dependency theorists, could only spell further disaster for less developed economies.

Dependency theory

Dependency theory rose to prominence in the late 1960s and early 1970s in the wake of disillusion about the prospects of economic growth in Third World countries. It provided a radical departure from the precepts of the modernisation school. While the latter had stressed that one way of aiding the development process was to expose the Third World to the influences of the West, dependency theorists argued that this exposure itself was highly damaging and frustrated the potential for self-determined growth. Growth in the West, the 'core' of the world economy, was seen to be achieved at the expense of exploitation of the Third World 'periphery'. As such, development and underdevelopment were two sides of the same coin.

Dependency theory suggested that contemporary relationships had their roots in past periods of merchant capitalism and colonial rule, when Third World areas had been pressed into specialising in the production of primary export products to suit the needs of the imperial powers. Those areas which had had closest links with the advanced economies were those most likely to be underdeveloped today. Despite the post-war efforts of Third World countries to diversify their economies, and in particular to promote industrial-isation, the continued strength, dominance and intervention of the West meant that the surplus value created by the labour of 'satellite' Third World nations was transferred (in the form of trade, interest payments and so on) to the advanced economies, thereby preventing any significant generation of wealth in less developed countries. This pessimistic scenario was often borne out in practice. For example, the 'import substitution industrialisation' model followed by major Third World regions such as Latin America (with the objective of cutting reliance on imports from the West and generating self-sufficiency) almost universally failed to produce the desired results.

Dependency theory not only drew attention to the way in which advanced economies might restrict the development potential of Third World nations, but also identified the role played by national elites in this process. The richer urban classes in Third World countries were seen to act as intermediaries between the capitalist core and the underdeveloped periphery. There was thus a chain of dependence from rural satellites of Third World countries, via the cities and elites of the developing periphery, to the capitalist centres of the advanced economies. But, although dependency theory offers more insights into the reasons for underdevelopment than modern-isation theory, it has several problems.

First, the idea that the world consists of two major groups, the 'exploiters' and the 'exploited', based primarily on money flows and outflows, is far too simplistic. Many have argued that it is not just a case of how much surplus is extracted from Third World nations, but also of the way in which that surplus is extracted. In this respect, it is vitally important to take account of the class structures of Third World countries as well as their external relationships with advanced economies. The role of the internal elite is argued by many to be just as much to blame as external dependency for creating widespread poverty and social inequality.

A second inadequacy of dependency theory is its argument that underdevelopment would be most acute in areas of greatest involvement with the advanced economies. In practice, places such as Mexico City and São Paulo, with long exposure to the West and with a good deal of multinational investment, turned out to be two of the richest cities in the Third World.

A third problem with dependency theory is that it offered little in the way of realistic policy measures. Dependency theory argues for a complete break with the capitalist metropolises, for socialist revolution and for collective self-reliance between Third World nations. Not only are each of these objectives extremely difficult to achieve, but in terms of people's everyday lives they can often cause serious upheaval and hardship (viz. the situation in Ethiopia today). Furthermore, many socialist countries such as Cuba merely exchanged one form of dependence (trade with the USA) for another (trade and aid from the Soviet Union).

The 'alternative development' and 'basic needs' approaches

As a response to the inadequacies of the policy prescriptions of both modernisation theory and dependency theory, a third set of ideas about development emerged in the 1970s, often referred to as the 'alternative development' approach. Alternative development is concerned less with explaining how contemporary patterns of global inequality have emerged than with policies for doing something about them. The main criticism of most development approaches has been that large scale, overly technological programmes with an industrial bias are totally inappropriate for the needs and resources of the Third World. In its place have been advocated approaches relying on small-scale self-sufficiency. This allows a much greater emphasis on equitable development, indigenous techniques and organisation, and lower levels of environmental disturbance.

Closely associated with alternative development thinking has been a recent body of thought referred to as the 'basic needs approach'. This urges that the primary goal of development should be the eradication of poverty and the fulfilment of basic needs. The question of what is meant by basic needs is open to debate but, generally speaking, it is thought that as an absolute minimum everyone should have guarantees of certain material resources such as food, fuel and shelter, and social resources such as health and education. Beyond these, other items which may enter a basic needs agenda include access to employment and political freedom.

Common to the alternative development and basic needs approaches is the idea that development strategies should be 'bottom up'; in other words that they should build upon the economic and social traditions of Third World people and be sensitive to their needs. This concept is diametrically opposed to previous approaches, whereby development projects have been 'top down', i.e. originating from national and international agencies and elites which often have little concern for traditional life styles or about whether particular forms of development are genuinely desired by the poor.

In the UN Second Development Decade (1970–80) there was a clear movement by the World Bank, the International Labour Office and other key agencies towards the basic needs approach. This emphasised the goal of development as the improvement of 'welfare for all', rather than economic growth for the benefit of only a few. Thus 'development' as a concept has begun to escape the definitional straitjacket of earlier modes of thought and is becoming a broader concept, more sensitive to local needs.

AGENTS IN THE DEVELOPMENT PROCESS

From the preceding accounts of approaches which have dominated development thinking over the last thirty to forty years, one thing is clear: an emphasis on simplistic ideas of Western models of economic growth does not lead to 'development' in Third World countries. But the various efforts and approaches have led to the growth of a number of means of institutional intervention. This intervention is undertaken by national governments, international development agencies or non-governmental organisations, and is frequently a critical determinant of whether development efforts occur and what form they take. A brief summary is provided below of the major agents in the development processes.

Third World governments

The types of policies implemented by national governments obviously vary according to the political regimes involved. Socialist governments and long-standing centralised authoritarian and/or military states are usually more likely to undertake long term plans than those countries with an active party system and/or where turnover of government personnel is more frequent.

Whatever the resources and political leanings of national governments, most are concerned first and foremost with large scale infrastructure development schemes, particularly the building of roads, dams, communications and energy supplies which facilitate economic development. They also oversee the growth of the economy via legislation on tariffs, exchange rates and industrial subsidies. Sometimes national governments group on a regional basis for the purposes of trade and investment, examples here being the Inter-American Development Bank and the African Development Bank.

International development agencies

International aid and development agencies wield an important degree of power over Third World governments in two main ways. First, they are often a critical source of funding and as such may determine the development path along which a Third World country proceeds. Second, they are frequently important in influencing the general direction of global development objectives and strategies.

The two most important international institutions responsible for providing financial aid to Third World governments are the International Monetary Fund (IMF) and the World Bank. Both were set up at the Bretton Woods conference in 1944 and today provide major sources of funding to Third World countries. IMF loans are generally deployed on a short term basis to solve immediate balance of payments problems, whereas World Bank money is generally used for longer term development projects such as the building of roads, schools and houses or the creation of employment. Both these organisations are heavily dominated and funded by the United States and do not have representatives from several centrally planned economies such as the USSR and various countries in Eastern Europe.

171

Beyond the IMF and the World Bank, the United Nations, whose membership is drawn from a far more representative number of countries (154 out of a world total of 162), is also very powerful and has long been concerned with the social consequences of development. To this end it has a number of agencies specifically concerned with social and welfare issues, including the Fund for Population Activities, the Children's Fund, and the Centre for Human Settlements and Housing. The United Nations plays a critical role in shaping global social policy initiatives and also attempts to promote international discussion and collaboration by sponsoring conferences and publications on themes designated as priorities for development. Recent examples here include the United Nations Decade for Women (1975–85) and the International Year of Shelter for the Homeless (1987). Similar international collaboration is stimulated by organisations such as the World Health Organisation and the International Labour Office.

Government development organisations in advanced economies

In addition to regional and international agencies, several Western governments have their own development organisations. These are usually concerned with promoting bilateral trade and other agreements with Third World states, as well as providing finance for development projects. Undoubtedly one of the most powerful organisations of this type is the United States' Agency for International Development, but other major organisations include the Swedish International Development Authority and the Danish International Development Agency.

Non-governmental organisations

Although on a far smaller scale than international development agencies, NGOs often play a major role in Third World development. Because they rely on private sources of income and are not tied to government policy, they are generally freer to experiment in the choice of development projects. Usually NGOs are much more committed than larger organisations to reaching the poorest of the poor, to responding to local needs and initiatives (i.e. to pursuing 'bottom up' development), to encouraging the participation of the beneficiaries in project implementation and to help Third World communities help themselves.

Some NGOs are sponsored by Church organisations such as Christian Aid, the Catholic Institute for International Relations, the Catholic Agéncy for Overseas Development and the National Christian Council of Kenya. However, others, such as Oxfam and Voluntary Service Overseas, have no religious affiliation.

The methods of different NGOs vary widely. For example, VSO responds to direct requests from Third World governments for personnel trained in specific skills such as carpentry, nursing or administration who in turn train Third World people to do the job themselves. Christian Aid, on the other hand, tends to provide funding rather than personnel.

CONCLUSION

Although economic growth is still a primary concern of Third World countries, and many still believe that the wealth generated by growth in modern industrial production and trade will in time spread to all sections of society, past experience has shown that even dramatic and sustained improvements in GNP performance have by no means guaranteed the extension of benefits to the Third World poor. If 'development with equity', along the lines suggested by the advocates of the alternative development and basic needs strategies, is to become a realistic prospect, then major changes in the handling of development planning are called for.

Such planning must take into account the fact that the Third World is not a uniform place. The development process must recognise the particular characteristics and problems of each area, along with the needs of particular social groups within those areas. More detailed examination of particular Third World problems, and a range of policy responses, is undertaken in the following chapter.

QUESTIONS

1 What are the main characteristics and problems of modernisation theory?
2 Why is dependency theory inadequate as a guide to policy?
3 What are the characteristics of the alternative development and basic needs approaches?
4 Who are the chief agents in Third World development and how has their role changed in recent years?

FURTHER READING

Crow, B. *et al.* (1986) *The Third World Atlas* (Open University Press: Milton Keynes). Accessible introduction to global inequalities through a series of maps and concise interpretations.

Hayter, T. and Watson, C. (1985) *Aid: Rhetoric and Reality* (Pluto: London). Penetrating analysis of the structure and operation of major international institutions in Third World development, particularly the IMF and the World Bank.

Hulme, D. and Turner, M. (1990) *Sociology and Development: Theories, Policies and Practices* (Harvester Wheatsheaf: London). A broad and comprehensive source for the study of Third World development, including analysis of measures of assessing development theories, planning approaches and case study examples.

Roxborough, I. (1979) *Theories of Underdevelopment* (Macmillan: London). Excellent review of development theories with extensive reference to actual experiences of development strategies in various Third World countries.

Stewart, F. (1985) *Planning to Meet Basic Needs* (Macmillan: London). One of the most detailed accounts of Basic Needs approaches to development, including case studies of attempts to fulfil Basic Needs objectives in a range of Third World countries.

Sachs, W. (1993) (ed.) *The Development Dictionary* (Zed Books: London). An often controversial overview with specific discussions of environment, technology and equality.

9

NATIONAL PERSPECTIVES ON THIRD WORLD DEVELOPMENT

Sylvia Chant

The previous chapter discussed various aspects of the global problem of the Third World. In this chapter greater attention will be paid to the variety of Third World conditions, and the geography of development, through examination of three interrelated issues: population (including the urban question), employment and welfare provision.

THE GEOGRAPHY OF POPULATION

One major factor distinguishing Third World countries from advanced and centrally planned economies is their high rate of population growth. While the average annual rate of growth in world population was estimated at 1·7 per cent between 1980 and 1985, that of more developed regions was under 0·7 per cent, compared with over 2·0 per cent in the less developed regions (Table 9.1).

Although, at a general level, population growth in the Third World has tended to decrease slightly since the early 1970s, in certain areas, notably Africa, rates are still rising. By the turn of the century the global population will probably be about 6,100 million and the Third World will have increased its share to nearly 80 per cent, from just under 75 per cent in 1980.

Massive population increase in Third World countries is primarily due to the fact that crude birth rates have remained far higher than crude death rates, even though both have declined since the 1950s (Figure 9.1). Life expectancy at birth in less developed countries rose from forty-two years in 1950–55 to fifty-five years in the period 1975–80. By 1988 it had climbed to sixty-two years.

Table 9.1 Global and regional population indices

Region	Average annual rate of increase (%)			Total population 1980 (million)	Crude birth rate 1975–80 (per 1,000)	Crude death rate 1975–80 (per 1,000)
	1970–75	1975–80	1980–85[1]			
World	1·91	1·72	1·70	4,432	28·5	11·4
More developed regions	0·84	0·71	0·68	1,131	15·8	9·4
Europe	0·63	0·40	0·34	484	14·4	10·5
Northern America	0·86	0·86	0·95	248	16·3	9·1
USSR	0·95	0·93	0·93	265	18·3	9·0
Japan	1·33	0·88	0·62	117	15·1	6·3
Australia–New Zealand	1·68	1·21	1·15	18	16·8	7·9
Less developed regions	2·32	2·08	2·04	3,301	33·0	12·1
Africa	2·73	2·90	3·00	470	46·0	17·2
Latin America and Caribbean	2·54	2·45	2·38	364	33·6	8·9
East Asia (excl. China and Japan)	2·21	2·05	1·92	63	27·1	7·9
South Asia	2·36	2·22	2·17	1,404	37·1	14·8
Melanesia	2·73	2·70	2·72	4	42·3	15·1
Micronesia	1·98	1·74	1·66	1	34·0	7·0

Note: [1] Estimate.
Source: United Nations World Population Prospects as assessed in 1980, United Nations: New York, 1981

However, high birth rates mean that young people still form the bulk of the population, and this has major implications for continued high fertility.

Population has not only grown massively in the post-war period, but has also become markedly more urban. In Zambia, for example, only 23 per cent of the population was urban in 1965, but the share rose to 54 per cent by 1988. In India the total urban population increased by 38 per cent in the decade 1961–71 alone.

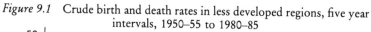

Figure 9.1 Crude birth and death rates in less developed regions, five year intervals, 1950–55 to 1980–85

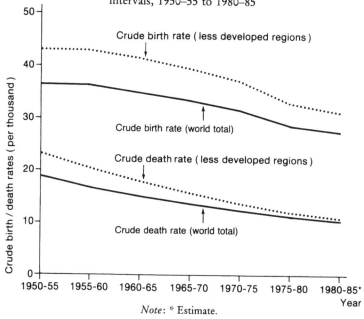

Note: * Estimate.

Urban population growth rates since the 1940s in the Third World have been between 3 and 5 per cent per annum, well above national growth rates. Although natural increase (the growth of the population *in situ* as a result of an excess of births over deaths) is now the primary motor of demographic growth in populous Third World cities such as Mexico City and São Paulo, migration has generally been the principal factor in post-war urbanisation.

Migration

There are several different types of migration, which often selectively affect different groups of the population, and vary considerably both between and within various Third World regions. Migration to towns and cities is usually linked with an urban bias in capital investment and employment opportunities.

In geographical terms, migrants tend to move from poor lagging regions towards relatively wealthy urban industrial areas, although this observation is by no means universal. Some authors argue that out-migration is far more likely from richer than from poorer agricultural regions. This is because people are more likely to lose their livelihood as a result of the concentration of land or the mechanisation of agriculture in richer farming areas than in those characterised by subsistence farming on small plots. Furthermore, since the most developed agricultural regions are likely to be nearest to large cities, migration is facilitated through improved transport links, exposure to information, and so on.

Another important qualification is that migration to urban areas is not always permanent. Until recently in East Africa, for example, there was a tendency to *circular* migration, where young rural men would move to the towns to work, but return to their families once they had accumulated sufficient capital to buy their own land. Migration may also be *seasonal*. In Mexican tourist towns, for example, men often find jobs as waiters, barmen or hotel porters during the high season, but leave their families in town and return to rural areas during the rainy summer months to farm their own plots, to hire themselves out as day labourers on large farms, and sometimes to migrate illegally to the United States. One factor is clear, however, rural–urban differentials in employment opportunities, wages, and access to social resources such as education and medical care, play a major part in prompting cityward migration.

The question of who moves to cities also depends on *time* and *education*. In the early stages of migration it is often the younger, better educated members of rural communities who move. Youth is not only usually associated with freedom from marital and/or parental responsibilities but, along with a basic educational qualification, is often a critical passport to an urban job. Once young migrants have established themselves in the city it is highly likely that their relatives will join them. As a result, migrant selectivity on the basis of age and socio-economic status diminishes over time: first, because the pool of potential migrants in rural areas grows smaller and, second, because later migrants with perhaps less education can rely upon support from kin in finding work and shelter.

Migrant selectivity also varies according to *gender*. In some parts of the world such as Latin America and South East Asia, where various industries (multinational assembly plants) and services

Plate 9.1   Rural scene: the village of Montezuma, Costa Rica
(*Photo*: Sylvia Chant)

(domestic service, the sex trade) recruit significant numbers of
women, more women tend to move to cities than men. In the case
of Latin America this is also related to the fact that women's
job opportunities in agriculture are very low (in most Latin
American countries less than 10 per cent of the agricultural
labour force is female), and thus they are forced to move to
the towns to find work. In Africa, the Middle East and South
Asia, by contrast, men tend to dominate urban migration flows.
In Africa this is partly due to the fact the women often form
a large part of the agricultural labour force (over 40 per
cent) and therefore have less need to move to cities. In South
Asia and the Middle East it is also related to the traditional
control by men of women's spatial mobility. For cultural reasons,
women are rarely allowed by their families to move alone to
cities.

Urban primacy and 'hyperurbanisation'

Until very recently most migration in Third World countries has been to major cities because they contain the bulk of investment and employment opportunities. In 1980, for example, 47 per cent of all manufacturing employment in Mexico was concentrated in the capital (from a level of 27 per cent in 1950). Currently Mexico City has a population of about 19 million people (approximately a quarter of the entire national population and around one-third of the urban population) and is over four times larger than Mexico's second largest city, Guadalajara.

This pattern of *urban primacy* (where the population of the first city of a country is substantially more than double that of the second city, more than three times that of the third, and so on) is repeated in many other Third World countries. For example, data from the early 1980s suggested that Luanda was ten times as large as Lobita and Benguela, the second and third Angolan

Plate 9.2 Urban scene: San José, Costa Rica
(*Photo*: Sylvia Chant)

towns; Kingston, Jamaica, twelve times larger than the second settlement, Montego Bay; and Bangkok forty times larger than Chieng Mai, the second city of Thailand. In very small countries such as Costa Rica, where primacy is far more likely because the whole country is an urban hinterland, nearly two-thirds of the urban population and one-third of the national population reside in the San José metropolitan region. If other cities in the Central Valley such as Alajuela, Heredia, Cartago and Turrialba are included, nearly three-quarters of the Costa Rican urban population is contained within a 40 km radius of San José (see Table 9.2 and Figure 9.2).

Table 9.2 Population of major towns, Costa Rica, 1984

Town	Population
Total population	2,416,809
Urban population	1,075,254
San José (metropolitan region)	635,191
Alajuela	42,786
Heredia	41,439
Cartago	33,962
Limón	33,925
Puntarenas	28,390
San Isidro	14,529
Turrialba	13,898
Liberia	12,335
Ciudad Quesada	11,119

Source: *Census of Population 1984*, Vol. 1, San José, Costa Rica, 1987

Third World urbanisation is a question not just of relative size, however, but also of absolute size. Several cities, including Bombay, Calcutta, Cairo, Rio de Janeiro, São Paulo and Buenos Aires had, in 1985, populations of 5 million or more (Table 9.3). Such centres are often labelled 'hyperurbanised', giving the impression that they have grown not only too rapidly, but also out of all proportion to their capacity to provide jobs and housing.

Primacy and hyperurbanisation are frequently attributed to two major causes; first, previous patterns of colonial rule whereby administrative authority and economic dependence were imposed upon one or two major centres, leading to a self-perpetuating process of concentration of investment and markets and, second, contemporary development policy, which favours certain areas over others.

Figure 9.2 The location of major towns, Costa Rica

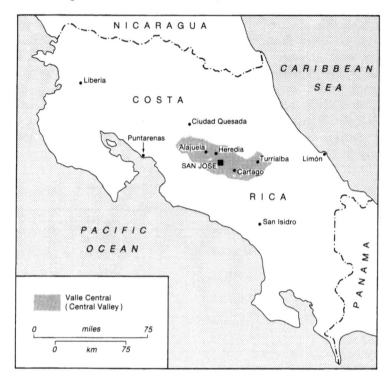

Although primacy and hyperurbanisation are not necessarily a 'bad thing', and indeed are not limited to Third World countries, there are probably three major negative implications of such concentration. One is a problem of regional imbalance. While resources and markets remain concentrated in cities, it is unlikely that rural areas will benefit from development. Another is a problem of scale itself – many cities not only have massive populations at present, but their populations could well double by the year 2000. A final, related, problem is that of management; how can urban planning authorities cope with such rapid, spontaneous expansion? In the following section we briefly consider some of the major policy responses to demographic growth and distribution in the post-war period.

Table 9.3 Largest Third World cities

City	Population of urban agglomeration[1]	Share of national population (%)[2]
Mexico City	14,750,182	18·8
São Paulo	10,099,086	7·4
Buenos Aires	9,967,826	32·6
Calcutta	9,194,018	1·2
Seoul	8,364,379	20·2
Bombay	8,243,405	1·1
Manila	6,720,050	12·3
Cairo	5,875,000	12·1
Tehran	5,734,000	12·9
Delhi	5,729,283	0·8
Rio de Janeiro	5,615,149	4·1

Notes: [1] Figures are for 1985 or for the most recent year before 1985.
[2] Mid-year estimate, 1985.
Source: United Nations Demographic Yearbook 1985, United Nations: New York, 1987

Policy responses to population problems

With regard to population growth *per se*, the overriding conclusion of the World Population Conference at Bucharest in 1974 was that if the quality of life was to improve for Third World nations, then population growth had to be reduced, and to be reduced drastically. The principal solution offered was population control, and specifically the introduction of family planning.

Since the mid-1970s several Third World governments, backed up by substantial foreign aid, have established a variety of population control programmes. At one end of the spectrum this has sometimes resulted in compulsory sterilisation of Third World men and women. At the other it has involved more 'voluntary' forms of persuasion to adopt Western contraceptive practices, often operating on a system of rewards and incentives – radios for vasectomies in India, farm animals for birth spacing in Thailand, and priority in employment, housing and education for small families in Singapore.

Countries with very active 'voluntary' family planning campaigns include Thailand, Mexico and Colombia, which between the 1960s and 1980s have succeeded in bringing down their birth rates by approximately one-third and, through a variety of means, have persuaded around 50 per cent of women to use contraceptives on a regular basis.

However, population control is a very controversial issue.

183

Contraceptives promoted in many Third World countries are often outlawed in the West because of dangerous side-effects, yet dumped cheaply on Third World markets by major drug companies. Furthermore, many would argue that the West is concerned to keep numbers down only because it fears the repercussions of social and political unrest on the world economic system.

While population control has been urged as the panacea for growth, redistribution has been the prescribed cure for concentration. There are two principal policies for diverting growth away from major cities. The first includes trying to keep the population where it is via investing in rural development, or actually prohibiting people from migrating to certain cities; the second involves diverting migratory flows away from major urban areas into smaller 'secondary' or 'intermediate' centres. This second policy is known as *decentralisation* and has been pursued with some success in countries such as Malaysia and Mexico. In the latter, incentives are offered to industries to set up in intermediate centres. Sometimes decentralisation also involves the creation of new towns in lagging regions, whether as major new industrial growth poles, such as Ciudad Guayana in Venezuela, or as new administrative centres such as Brasilia, the capital of Brazil.

THE GEOGRAPHY OF THIRD WORLD EMPLOYMENT

In the preceding section it was noted that migration is more often than not a response to real or perceived spatial differentials in employment opportunities. Unfortunately, for those who are forced off the land or who choose to move to cities, waged employment is sometimes just as difficult to find in the cities as it is in the countryside.

In both rural and urban areas, economic development has often led to extreme shortages of jobs and/or opportunities to generate income. In rural areas of the Third World, for example, progress in agricultural production, involving the concentration of land, mechanisation and the introduction of high-yielding Green Revolution crop varieties, has frequently been achieved at the expense of the livelihoods of the rural poor. In India and Pakistan, poor tenant farmers have often been evicted to allow landlords to undertake expensive irrigation projects. In Central America the massive agro-export boom in sugar, beef and cotton since 1950 has critically reduced opportunities for subsistence farming and agricultural employment, and prompted a situation where rural labour must constantly shift around from harvest to harvest, or move permanently to towns.

In urban areas the story is much the same. Large scale capital-intensive manufacturing excludes many of the migrant poor from regulated waged employment and, in the absence of state welfare benefits, large numbers of people live in extreme poverty.

Economic dualism

In the countryside, large scale mechanised export-oriented agriculture stands alongside, and often in direct conflict with, the small scale subsistence sector. In towns, only a small fraction of the population are employed in 'modern' industries, registered commercial enterprises and public services; the rest have to resort to casual labour, self-employment and, sometimes, to 'criminal' activities such as petty theft and prostitution. The coexistence of two broad sectors of employment in both rural and urban areas of the Third World has given rise to the notion of 'economic dualism' and to the application of the terms 'informal' and 'formal' sector to these different kinds of income-generating activities. Table 9.4 summarises some of the typical characteristics of each sector, although it should be noted that there are not only considerable variations within them, but also several links between them.

Proximity to markets probably makes it slightly easier for people

Table 9.4 Characteristics of dualism in Third World employment

Formal sector	Informal sector
Large scale	Small scale
Modern	Traditional
Corporate ownership	Family/individual ownership
Capital-intensive	Labour-intensive
Imported technology/inputs	Indigenous technology/inputs
Protected by labour legislation	Absence of formal labour legislation
Difficult entry	Ease of entry
Formally acquired skills	Informally acquired skills
(e.g. school education/training)	(e.g. craft apprenticeship)
Capitalist mode of production	Peasant mode of production
profit-oriented	subsistence-oriented

Source: Adapted from David Drakakis-Smith, *The Third World City*, Methuen: London, 1987, table 5.5 p. 65

in towns to survive from informal employment than those in remote rural regions. Indeed, a wide range of activities may be undertaken, including traditional marketing, street hawking, petty commerce, the home production of foodstuffs and cheap consumer durables such as shoes and household goods. The income generated by many of these activities is rarely included in GDP, even though it supports a large proportion of the population.

Despite a persistent association of the informal sector with un- or underemployment, in recent years attitudes towards it have gradually become more favourable. Many have argued that the informal sector represents an important source of indigenous national production, may produce a significant number of consumer goods and thereby reduce dependence upon large scale import substitution enterprises, is capable of higher growth rates than the formal sector and may absorb large amounts of labour because of its low reliance on capital-intensive technology.

Policy responses to employment problems

While the previous tendency of many governments had been to clamp down on informal activities, from 1970 onwards institutions such as the International Labour Office have recommended government support.

The wholesale transfer of productive investment to the small scale sector is obviously out of the question when so many Third World governments and elites have a vested interest in promoting capital-intensive production. However, it cannot be denied that the informal economy acts as a safety valve for outright destitution and social unrest. Were it not that the poor were able to scratch some kind of living (however basic) from informal activities, there would be no alternative but to provide huge amounts of unemployment benefit to massive numbers of people, which few Third World governments are able or willing to do. Until now, only a handful of developing countries, such as Tanzania, have adopted a major policy of small scale economic self-reliance (in this case known as 'rural socialism'), but in several areas governments have consented to the removal of various forms of restrictive legislation on informal employment, and in some cases have provided credit to petty entrepreneurs and small businesses. International development agencies and NGOs have also become involved in sponsoring small scale 'bottom up' production and employment projects in both rural and urban areas.

An example here is the income-generating project set up in Dev-Dholera, a small village in the state of Gujerat, India. A few years ago the village was hit by drought, leaving most of the population poor and hungry. Men were forced to migrate to the nearby town of Ahmedabad to find work, but were often unsuccessful, and there was little money to support their families. A weaving and caning co-operative set up in the village in 1984 by the Self-employed Women's Association (SEWA) has provided a partial answer to Dev-Dholera's problems by commercialising the traditional skills of a small group of female community members, and extending them to others.

SEWA was originally founded in 1972, with financial assistance from the United Nations Development Fund for Women, to assist women workers in the informal sector with loans, training and management skills. At present it has 50,000 affiliates. Although most of SEWA's activities are concentrated in urban areas, in 1977 a rural wing was founded to help rural women, thereby benefiting villages such as Dev-Dholera.

Prior to the establishment of the co-operative there were only three weavers in Dev-Dholera; now ten trained weavers and a total of thirty illiterate girls and women between the ages of sixteen and thirty work from 11 a.m. to 5 p.m. at the co-operative. Young apprentices earn eleven rupees a month (US$0·93) and after training receive a weekly average of about forty-two rupees (US$3·44) in return for three small dhurrie carpets. The carpets are then sold by SEWA in Ahmedabad for fifty-five rupees (US$4·50) each, with the bulk of the profits being reinvested in further looms, cottons and dyes for the co-operative.

Other activities in Dev-Dholera now include training in caning, pottery-making and tailoring, and a dairy co-operative. The hours worked by women in these projects are deliberately limited so that they have time to do their domestic labour and help their husbands in the fields. This initiative, while small scale, has helped to stave off destitution for several families and has also provided women with an independent source of income. It also testifies to the fact that financial support can turn traditional skills to economic advantage.

However, while support for the informal sector can often mean an immediate material improvement in the lives of low income people, it is important to remember that its subordinate position to the formal economy means that the assistance provided may in the end be more favourable to those outside it. Several studies have noted how the informal sector benefits the dominant classes through providing cheap components, labour and retail services. It also heads off social unrest through absorbing people who would otherwise be unemployed. As long as the informal sector remains in the shadow of the formal sector, support is likely to produce only comparatively limited and short term gains for low income workers. Real improvement in the livelihoods of the poor in less developed countries can probably only be achieved through a radical restructuring of the economy.

THE GEOGRAPHY OF WELFARE

Given inadequacies in employment in less developed countries it is hardly surprising that welfare provision also falls far short of people's needs. Here a brief résumé of problems in health care and housing is offered.

Health

In many Third World countries health care is low on the list of government priorities and reflected in low levels of public expenditure. Many spend less than 5 per cent of central expenditure on health care, compared with an average of above 11 per cent in industrial market economies. Furthermore, in many countries the share is now declining. The World Bank group of low income economies (all those with a per capita GNP of less than US$500 per annum, excluding China and India) allocated only 2·8 per cent of total government expenditure to health in 1988, compared with an average of 5·5 per cent in 1972. Defence, economic services and general government administration take the lion's share of public expenditure in most countries. Why health should be a low priority is not hard to understand. When governments are concerned with increasing economic growth, scarce resources are ploughed into production not welfare. Medical care is not usually regarded as a productive investment, even though many would argue that a healthy population is not only a goal of development but also a prerequisite.

Low relative expenditure on health care in the Third World means very low absolute expenditure as well, and translates into woefully limited provision of medical facilities. In terms of doctors per head of population, for example, in less developed countries the ratio of population per physician is on average 4,790 to one, and in some cases over 50,000 to one, compared with an average of 470 to one in advanced economies. This is reflected in low levels of health among the population. Average life expectancy in the Third World in 1988, for example, was fourteen years less than in high income economies. Infant mortality rates (the number of children under one year dying per 1,000 children in the same age group) were on average at least seven times higher in the Third World compared with the advanced economies. High rates of infant and child mortality stem from low nutritional intake and insanitary living conditions. Diseases such as poliomyelitis, gastroenteritis, pneumonia, bronchitis, tuberculosis, malaria and dengue fever are rife in areas of overcrowded, poorly ventilated housing where there is inadequate disposal of human excrement and where water supplies (if any) are unclean.

One of the basic problems with health care in Third World countries, over and above insufficient provision, is that it is also of an inappropriate type. Much government expenditure goes into large and costly hospitals, sophisticated equipment and imported pharmaceuticals, which are beyond the means and requirements of most of the population. Instead of highly technical, centralised facilities, people need basic medical care in the form of local health centres, immunisation schemes and elementary training in hygiene, nutrition and first aid.

Policy responses to health problems

Due to the obvious mismatch between health service supply and demand in Third World countries, the World Health Organisation in 1975 launched a major campaign of 'health for all by the year 2000' and argued for the allocation of finance to primary health care. This approach advocates the decentralisation of health services and the participation of communities themselves in basic medical provision, incorporating, as far as possible, traditional methods (e.g. natural herbal remedies) and local personnel (e.g. 'barefoot' doctors, 'traditional' midwives). Again, as with recent policies towards the informal sector, this represents a 'bottom up' orientation, sensitive to the customs of Third World people. Also intrinsic to primary health care is an emphasis on prevention rather than cure. This involves integrated health projects encompassing improvements in housing and basic community infrastructure, such as water and sanitation. The WHO has been fairly successful in promoting primary health care in developing countries, and this in turn has resulted in a variety of schemes, often sponsored by other international development organisations and NGOs. One such scheme includes government campaigns for mothers to breastfeed their children. Others include integrated community development projects such as those which have taken place in Manila, where 35 per cent of the population live in slums and squatter settlements.

In Barangay 865, for example, in the Pandacan area of the city, there were very high rates of ill-health, mainly due to the fact that the railway line running through the centre of the settlement was used as an open toilet. Since 1981, however, problems have been reduced dramatically, thanks largely to a sanitation project implemented by the Metro Manila Infrastructure, Utilities and Engineering Programme, with financial support from the World Bank. In addition, the Manila City Authority now pays workers from within the community to keep the railway track clean.

Other primary health care programmes operating in the Pandacan area (and reaching a total of 10,000 people in four settlements, including Barangay 865) are antenatal care, child immunisation schemes, monthly weighing sessions for babies and the training of local health workers. In a period as short as two years, malnutrition and illness have been reduced by 75 per cent, and far fewer people now have to seek medical treatment for diarrhoea.

However, no matter to what degree health care is restructured, health itself is often dependent upon nutrition. In countries such as Zaire, Chad and Ghana people's average daily calorie supply is only two-thirds of the estimated requirement, compared with over one-third in most advanced economies. In future, ensuring an adequate diet will have to become part and parcel of a realistic preventive health care package.

Housing

From the above it can be seen that health is often highly dependent upon the kinds of conditions under which people live. The vast majority of the poor in both rural and urban areas have inadequate housing and lack many basic urban services such as water, electricity, sanitation, rubbish collection and paved roads. In most major Third World cities at least one-third of the population live in slums.

Again, housing usually occupies a low rank in government investment priorities, and few states have embarked on fully comprehensive public housing programmes. Thus, just as in the case of employment, the poor have had to create their own shelter on land acquired illegally (either through squatting or unauthorised transaction), and out of whatever materials they can afford or find to hand (see plates). This kind of housing is normally referred to as 'self-help' housing because its construction depends on individuals rather than the state. The settlements are known as 'irregular' because initial occupation of the land is usually illegal and residents may have to wait several years for legal tenure.

The land occupied by irregular settlements is generally on inhospitable terrain (precipitous hill slopes in Rio de Janeiro, the

Plate 9.3 Self-help housing, Querétaro, Mexico. Note here, and in Plate 9.4, the variety of materials used, including corrugated cardboard, metal sheeting and the sides of fruit boxes (*Photo*: Sylvia Chant)

Plate 9.4 Self-help housing, Querétaro, Mexico (*Photo*: Sylvia Chant)

sides of canals in Bangkok, reclaimed swamps in Guayaquil, and so on), and often distant from transport, amenities and employment. Nevertheless, were it not for self-help housing, governments would have a huge homeless population on their hands. Therefore, in recent years, in line with policy shifts in employment and health, governments and international aid agencies have increasingly realised the wisdom of sponsoring self-help as a solution which makes the best use of abundant labour and scarce capital.

Housing policy responses

Since the early 1970s there have been two major types of support for self-help housing. The first consists of creating new self-help settlements, by providing people with a land plot, basic services and sometimes subsidised construction materials in order that they may build their own homes. These are known as 'sites and service' schemes and have been implemented in a number of countries, including Brazil, Zambia and Kenya. One of the most notable is the Dandora Housing Project, Nairobi. This was established in the mid-1970s by the Kenyan government, Nairobi City Council and the World Bank to cater for a massive housing deficit in the city. The aim was to provide a total of 6,000 serviced land plots over a five-phase implementation period lasting about five or six years. At later stages the project has received financial support from other organisations such as the United Nations Children's Fund and the Ford Foundation.

Dandora, which lies 10 km east of the city centre and covers a total of 350 ha, was selected as the project site because it was the largest tract of publicly owned land in the vicinity, adjacent to existing residential areas and within relatively easy reach of the city centre and two major industrial zones. Three types of plots were available in Dandora. 'Type A' plots ranged in size between 100 m^2 and 140 m^2, and were intended for the lowest income groups. These came complete with a 'wet core' of toilet and shower, and beneficiaries were expected to build one or two rooms in an eighteen-month period with the aid of a loan. During this time they were permitted to build a temporary shelter on the site which had to be demolished at the end of the building period. 'Type B' plots also measured between 100 m^2 and 140 m^2, but included a kitchen and store in addition to a wet core. Again a loan was provided for the construction of two further rooms within eighteen months, during which time the family were expected to use the kitchen as a shelter. 'Type C' plots measuring 160 m^2 and including a wet core, kitchen, store and one further room were sold at market rates in order to subsidise the sale of the other two plot types. Here no loans were given to beneficiaries.

Although Dandora has undoubtedly played a major part in helping to solve Nairobi's housing problem, there were several difficulties with the project. One was the fact that entry to the scheme required not only a down-payment of 600 Kenyan shillings (almost a year's wages for some informal sector workers), but also proof of a monthly income of between 280 and 560 shillings. This again was well beyond the wages earned by those in casual work, and as such the project was not accessible to the poorest of the poor. In order to gain access to the project many people, especially women heads of household, had to take out loans from informal moneylenders at usurious interest rates or borrow from relatives. Yet financial problems did not stop there. Zoning regulations prevented many households from carrying out business on their premises, thereby robbing many families of their livelihood, or forcing them to operate illegally. This considerably reduced their capacity to pay back construction loans and draws attention to the need for shelter programmes to make allowances for the economic as well as residential needs of low-income groups. (See Nimpuno-Parente, 1987, for a fuller discussion of the Dandora project.)

The other major type of self-help policy response is 'squatter upgrading'. This involves the improvement of the settlement *in situ* through the regularisation of land tenure, the provision of infrastructure and services, technical assistance and construction materials. These schemes are far more widespread than sites and service programmes because land is already acquired, which means lower costs. They also allow for genuine 'bottom up' development in the sense that they build upon what the poor have already achieved. Furthermore, if the neighbourhoods are long established, there is often some form of community organisation which helps to bring people together in co-operative labour. However, squatter upgrading is not without its problems. In the course of introducing services or laying roads some residents have to abandon their dwellings, while others have to sell up and move out if they cannot afford to pay for settlement improvements.

In the 1987 UN Year of Shelter for the Homeless the commitment to providing decent shelter for all, primarily in the form of self-help, was forcefully reiterated. However, it should be noted that land for housing is becoming increasingly scarce in many major Third World cities and self-help may become less viable. Evidence suggests that land shortages in many areas are resulting in an increase in renting and sharing (multiple family occupancy) in irregular settlements. This implies the need for new and more imaginative policy responses to housing problems in the future.

CONCLUSION

From the brief and selective account of development in the Third World in the past two chapters, it is obvious that prospects for dramatic improvement in the conditions of the poor are bleak. We have seen how, in the past decade or two, there has been a notable shift in the policies of governments, international aid agencies and NGOs towards supporting attempts by the poor to improve the quality of their lives by small scale, 'bottom up' projects in employment, health care and housing. Such projects appear to benefit some of the needier people in society and, perhaps, also help provide the groundwork for a more autonomous form of development. Nevertheless, efforts so far have been highly localised and piecemeal, and in themselves have achieved little in terms of redistributing national wealth. Furthermore, welfare programmes in many poor countries today face drastic cuts as the International Monetary Fund presses debtor countries into stringency in public expenditure, while a general slowdown in world trade means that the phenomenal growth rates of the miracle years of the 1960s and early 1970s are extremely unlikely to recur. The kinds of policies adopted from now on will depend very much on resources, as well as the long term objectives of Third World governments and international aid agencies.

QUESTIONS

1 How and why do birth and death rates differ between regions in Third World countries?
2 How does migration contribute to 'urban primacy' and 'hyper-urbanisation'?
3 What policies can be used to help employment growth in the Third World?
4 What are the problems of improving health care in the Third World and how can 'self-help' contribute to their resolution?

FURTHER READING

Brydon, L. and Chant, S. (1989) *Women in the Third World: Gender Issues in Rural and Urban Areas* (Edward Elgar: Aldershot). Concerned in general with women and poverty in the Third World, including discussions of population policies, employment, housing, health, welfare and a range of development projects discussed in the present chapter.

Gilbert, A. and Gugler, J. (1992) *Cities, Poverty and Development: Urbanisation in the Third World*, 2nd edition (Oxford University Press: Oxford). Informative account of Third World urbanisation covering employment, housing and migration. Lots of facts and figures.

Gugler, J. (ed.) (1988) *The Urbanisation of the Third World* (Oxford University Press: Oxford). Collection of papers by a range of experts in different aspects of Third World urbanisation, including discussions of population, migration, housing and employment.

Nimpuno-Parente, P. (1987) 'The struggle for shelter: women in a site and service project in Nairobi, Kenya', in Moser, C. and Peake, L. (eds.)

Women, Human Settlements and Housing (Tavistock: London). Detailed case study of World Bank-sponsored housing project in Nairobi with particular attention to issues affecting women.

Phillips, D. (1990) *Health and Health Care in the Third World* (Longman: Harlow). Excellent up-to-date review of health problems and health policies in the developing world, and the role of international agencies such as the World Health Organisation.

Standing, Guy and Tokman, Victor (eds) (1991) *Towards Social Adjustment: Labour Market Issues in Structural Adjustment* (ILO: Geneva). Wide-ranging geographical coverage which looks at the changing nature of employment in developing countries.

Allen, Tim and Thomas, Alan (eds) (1992) *Poverty and Development in the 1990s* (Oxford University Press in association with Open University Press: Oxford/Milton Keynes). An accessible series of accounts about a range of social, demographic and economic problems facing Third World countries.

10

DRAWING THE LINE BETWEEN TOWN AND COUNTRY

Michael Hebbert

Governments throughout the developed world expend much administrative effort drawing and enforcing boundary lines between town and country. The principal mechanism is the body of laws and techniques known as town planning, though laws regulating agricultural land tenure also play a part. This chapter focuses on town planning, taking the United Kingdom and Japan as case studies. Though the two cases differ greatly, they have in common a planning system which has the role of drawing and enforcing boundary lines between town and country. In terms of manpower and resources, this is the largest purely geographical management task undertaken by government. It is also one that probably generates more political friction than any other public mapping exercise. In contemporary Britain, particularly (though not exclusively) in the south-east, hardly a day goes by without some press coverage of controversy about countryside preservation, urban growth, and the planning system's role as umpire or adjudicator between the two. As such it has an important effect on the market for land. This chapter puts these planning decisions into an historical and comparative perspective. It asks why government has become involved in the operation of land markets around cities, what purposes government planning serves, and what consequences arise.

EARLY TOWNS

Some tasks of government, such as taxation, defence, criminal law and consumer protection, are as old as civilisation. Others – such as welfare policy – are new, and reflect the particular circumstances of

197

modern industrialised society. Town planning combines elements of both. From the dawn of civilisation we find rulers laying out streets, buildings and monuments for military and economic purposes and for their own greater glory. Conscious design of towns is more than a symptom of civilisation, it is one of its essential attributes. But the issue of separating town from country is a distinctively modern preoccupation of government.

Historically, the boundary line of most towns in most countries was a wall. Town walls transformed mere clusters of buildings into definite physical entities. Military logic heightened the barrier effect of the wall, dictating compact, high density settlement in an open setting: the more compact and densely built the town in relation to the number of inhabitants, the more defenders could man the ramparts in case of attack. The fewer buildings outside, the less cover for attackers. It was also a matter of sheer convenience to be on the right side of the wall. Urban fortifications such as Theodosius's walls of Constantinople (Figure 10.1) built 1,600 years ago, are obstacles, just as is a river or a mountain. These walls still defined part of the western boundary of the city of Constantinople (now called Istanbul) in recent times.

Although artillery was eventually to make fortifications and seiges redundant, its effect in the sixteenth to nineteenth centuries was to reinforce their importance. Earthworks, masonry defences and waterworks were combined in ever more elaborate and

Figure 10.1 The walls of Constantinople. Built by the emperor Theodosius (337–61), the great triple walls for centuries made an impassable barrier between the city and an often hostile countryside.

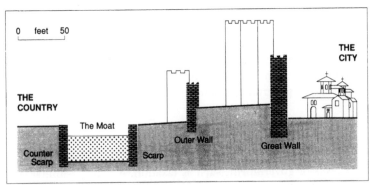

Source: C. Stewart, *A Prospect of Cities*, Longman, London, 1952

space-consuming patterns around strategically important settlements. Immediately outside the walls stretched an extensive glacis on which no building was allowed lest the defenders' line of fire was broken. Not until the third quarter of the nineteenth century was the principle of defensive fortification finally abandoned in mainland Europe, releasing a vast swathe of land for development. This was a windfall of potential development land comparable to that provided by inner city derelict land in the late twentieth century.

Whether or not towns were walled and gated, factors additional to defence combined to keep pre-modern towns as compact and distinct physical entities. First, most towns were legally distinct areas, with social organisations and political structures fundamentally unlike those outside. In both western and oriental feudalism we find town-based merchants using their command of ready money to buy a shell of relative autonomy from pervasive obligations to the landowning class:

> The mediaeval city was the classic type of the closed town, a self-sufficient unit, an exclusive Lilliputian native land. Crossing its ramparts was like crossing one of the still serious frontiers in the world today. You were free to thumb your nose at your neighbour from the other side of the barrier. He could not touch you.
> (E. Braudel, *Capitalism and Material Life 1400–1800*, New York: Harper & Row, 1967, pp. 402–3)

The rise of the modern nation state challenged the traditional rights and privileges of towns. They came to be absorbed within larger state-wide systems of law and commerce. As old self-governing jurisdictions were dissolved, towns began to acquire new meaning as symbols of a uniform order, expressed through a style of classical architecture. This style had been rediscovered in the Renaissance and was employed throughout the Western world until well into the nineteenth century. As the legal importance of town boundaries diminished, their architectural or cultural significance grew:

> The Renaissance town was a proud, sometimes even an arrogant, gesture of conscious triumph. The strong dramatic contrast of its frank artificiality with the 'natural' countryside was a purposeful and deliberate expression of the Renaissance

attitude of mind. And we have to remember how strong that contrast was. The Renaissance town made no acknowledgment of the countryside even at its edges. The suburb as we know it today did not exist. The edges of the town were as urban as the middle – often more urban. The London squares, those places which we now regard as essentially central-city in character, were originally built in open fields. So were the fine formal crescents and squares of Edinburgh New Town. So were those of Bath. ... To the Renaissance mind the town was the town and the country was the country, and though the twain must necessarily meet they should certainly not mingle. (Thomas Sharp, *Town Planning*, Harmondsworth: Penguin Books, 1941, p. 42)

Military, legal and architectural factors reinforced the other causes of compactness. The economics of urban life, in an era when movement was on hoof or foot, placed a high premium on close physical access. Each town was a self-contained employment area, in which almost everyone walked to do their business or practise their trade. If a city's population increased, its built-up area (walls and topography permitting) tended to expand evenly like ink on blotting-paper. As a result accessibility was maximised to all within the area of the town.

Urban regions were also largely self-contained in terms of food production. Around larger towns a girdle developed of intensive market gardening, dairying and stock rearing. This was often the laboratory in which improved agricultural practices were evolved, such as the system of rotation of crops or the introduction of new crops such as turnips. Later these changes diffused through the wider rural economy. These peripheral market gardens were enriched by the urine and faeces of the town dwellers (only recently piped away underground in Western Europe and still not in much of the developed Orient). The market gardens were highly productive and valuable properties whose ground rents could exceed those of land for building purposes. The nature of these two rent structures led, in early cities, to a solidly built-up area circled by a zone of intensive agricultural holdings (Figure 10.2). Thus the competitive sorting of uses in the urban land market (which we often call the 'ecology' of the city) reinforced once again the tendency towards physical containment. This gave the built-up area an unambiguous edge.

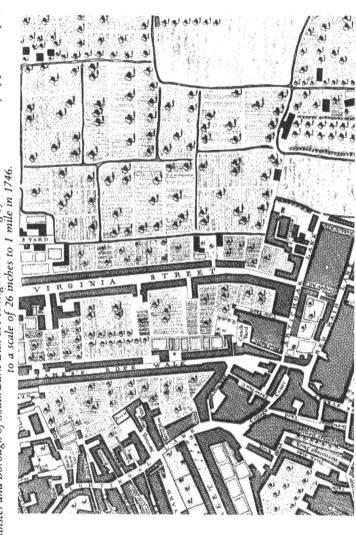

Figure 10.2 Market gardens on the fringe of Georgian London. A detail of Wapping, from the *Plan of the Cities of London and Westminster and Borough of Southwark with the contiguous Buildings, from an actual Survey by John Rocque, in 24 sheets to a scale of 26 inches to 1 mile in 1746.*

Source: Reproduction published in 1971 by Harry Margary, Lympne Castle, Kent

We may say, then, that a range of factors – military, legal, architectural and economic – conspired in history to define a more or less stable and tangible boundary between compact, continuously built-up towns and their hinterland. Other forms for towns were technically feasible. Very different towns occur in ancient Latin American civilisations. They had populations of hundreds of thousands settled over hundreds of square miles. But they had no counterparts in Europe.

TOWNS TODAY

Today, every one of the centripetal factors that historically encouraged compactness in urban areas has swung the opposite way. There is no need to labour this point: you can readily confirm it by comparison of city maps before and after the impact of industrialisation and modernisation. Military logic in an age of aerial warfare encourages dispersal of targets, not concentration. Legally, the status of citizenship is no longer distinct in the modern nation state and town boundaries offer no privileges. Economically, the development of transport and communication technology has weakened the pulling power of central towns, making it possible to live and to do business within a much broader geographical radius than the tight compass of walking distance. Cities no longer look to their immediate vicinity for food. Rising living standards have allowed a much larger consumption of living space per person. Modern domestic architecture reflects the individualistic preferences of consumers. Perhaps this is most evident in suburban houses and gardens.

From the mid-nineteenth century these distinctively modern forces began to work on the urban fabric, creating cities radically different from any seen previously. Cities grew rapidly in number, in population, and vastly larger in physical scale.

It is not the increase of urban populations that concerns us in this chapter, but the physical spread of the built-up areas. The spreading tendency of the modern city was first apparent in the frontier cities of early nineteenth century America. Here, land speculation, untrammelled by ownership constraints or existing structures, and facilitated by the predictability of a grid iron street layout, chased profit through extensive swathes of territory around any population nucleus. The resulting towns bore no resemblance to the dense population nuclei of old Europe. After the city's incorporation in

1802, Cincinnati's population of less than 1,000 occupied an urban area of 3·2 square miles – slightly larger than Amsterdam, with a population over 200,000. In mainland Europe, city walls maintained high urban densities for a further century. In Britain, however, in the nineteenth century, the centrifugal forces of land speculation soon began to produce a form of low density urban spread. This accelerated in the second half of the nineteenth century with the introduction of horse trams and electric trams and the branching of suburban railways. As one observer wrote, 'It was as though the compulsion in the earlier years of the century to bring the growing numbers into the cities and large towns had been replaced by a tendency to scatter them over more and more distant suburbs.'

It was the rise of free market capitalism that stimulated urbanisation on this unprecedented scale. It also endowed city life with new physical forms. It broke through the invisible membrane that had hitherto bound suburbs inseparably to their parent city. To understand why this should be, consider more closely the process by which the modern city grows. It is built, for the most part, by commercial developers who make their profit from buying farmland which has a low unit value, and selling it as urban land with a high unit value, sometimes equipped with services and buildings.

The biggest profits, around a growing urban nucleus, are to be made by those who first subdivide open farmland for the use of town dwellers, purchasing cheap and selling on the crest of the price rise as the area becomes built-up. Pursuing such gain, speculators seek out potential sites in anticipation of demand. This occurs first along existing radial roads, along railway lines, and along new roads. The speculative process is essentially prodigal with land. It scatters the urban influence far and wide, leaving much land unexploited behind the leading edge of development. Such land may often lie in an ecological limbo for years, being too valuable for farming but not yet ripe for urban development. In periods of boom, areas of thousands of square miles stretching to a radius of fifty miles around major cities may be subject to sporadic speculative development, often far in excess of demand. This occurred in California and Florida in the 1920s, around London in the 1930s and Buenos Aires in the 1960s. If the typical pre-modern city was a compact, dense settlement surrounded by a belt of market gardens, its modern counterpart has what is sometimes termed a 'formless form'.

The oil-slick spreading quality of the modern city, when left to its own devices, astonished nineteenth century observers. It was seen as a symbol of the disintegration of contemporary social life. Most people were disquietened by the voracious appetites of the modern city. They saw its growth in a negative light as destructive of both countryside and population. Other writers, such as Prince Kropotkin and the novelist H. G. Wells, painted rosier views of an emerging world in which the line between town and country had broken down, allowing industry and its workers to be scattered at low densities through the landscape. The car maker Henry Ford, who made an eminently practical contribution to the process in techniques of mass-production that brought car ownership within the reach of ordinary people, foresaw in 1922 that 'the ultimate solution will be the abolition of the City, its abandonment as a blunder. We shall solve the City problem by leaving the City.'

PLANNING TOWNS

What was the attitude of government towards the new phenomenon of urban expansion? Originally the growth of towns seemed to lie beyond public control. If landowners wanted to subdivide their land into building plots, that was their business, and their right. However, towards the end of the nineteenth century it was becoming apparent that dispersed urban development had significant costs. As a result there were calls for government to regulate the building process in the public interest. This led to 'town planning' legislation around the turn of the century in almost all the modern countries. In essence, government stepped in to restore the boundary line between town and country that no longer arose naturally.

We can group the considerations that encouraged government to intervene in the processes of city growth under three major headings.

The original, and still the most universal, motive for intervention is for efficient provision of what is called the *infrastructure* of a town. This is the engineering that sustains the visible architecture of cities. Built-up areas are more than buildings. Equally important is the equipment, much of it buried underground (hence 'infra'), which serves the separate buildings and enables their occupants to live modern urban lives: surfaced roads, fresh piped water, drainage and sewage systems, electricity, gas and telephone connections and weekly rubbish collections, as well as parks, schools, hospitals and fire stations. The cost of providing this infrastructure to new housing is high. It often equals the cost of the buildings themselves. The costs also vary with the physical pattern of development. A town without any planning, where speculators have been free to experiment, tends to be widely spread out and correspondingly more expensive to service with infrastructure than one which is compact. Here is a major stimulus to government to step in. On behalf of the general taxpayer, and in the interests of public economy, it defines a limit beyond which land may not be divided up for development, in order to concentrate infrastructure investment in the areas already developed.

The second reason why government has become involved in managing the demarcation of town and country is a desire to protect the *rural* economy from the influences of urban land speculation. Urbanisation is, practically speaking, an irrevocable process – once built up, land cannot usually revert to its role of food production. Unconstrained urban sprawl lays what has been called a 'shadow' over the countryside. Not only is more land built up but the buildings exert an unsettling effect over a wider area, encouraging speculative trading and land-holding and causing the running down of agricultural activity in expectation of land sale. In countries with a limited stock of cultivable land, strategic concern for the protection of food production has been an important stimulus to city planning controls. Besides, farmland may be valued and protected for other than agricultural reasons. In many north European countries, people see an intrinsic beauty in a green and open agricultural landscape, and expect government to protect it in the same way, and for comparable reasons, as the nation's archaeological, architectural and artistic heritage. Increasing environmental concern has reinforced pressure to protect rural and agricultural landscapes.

The third motive for intervention arises from *urban* policy of various kinds. Government may attempt to draw a line around cities from a desire to stabilise the social and economic structure of the city. It may be the behaviour of urban land markets that prompts intervention. Because of the great financial gains to be won, and the poor quality of information available to buyers and sellers (who are by definition dispersed), free speculative land markets are intrinsically unstable and liable to powerful surges that suck in savings unproductively from the economy at large. Japan's planning system was created in 1968 after a period of hyperinflation in land prices that reached an annual rate of 42 per cent in the single year of 1961. In other cases it has been the social and political dimensions of urban change that have prompted governmental action. A shifting urban frontier has been seen as a public risk for a very mixed bag of reasons. For example, in Britain the prosperous classes of society have tended to move out of cities, leaving the poor behind, so creating political, social and economic problems. In Third World cities the rural poor have moved into cities and created their own belts of suburban poverty. Either way, the demand arises for measures to stabilise the social structure of cities by restricting their physical growth.

TYPES OF PLANNING

For any of such reasons, singly or in combination, government may step in to draw a line between town and country. The principal mechanism used is the town plan, a map with legal force which defines areas for building and areas for protection or for public services. The degree of planning intervention varies widely from country to country, depending on the degree of public concern over infrastructure provision, farmland protection and urban growth. A tight planning system will define unambiguously the layout, contents and boundary of each successive urban extension, rolling the boundary line forward as necessary at each plan period. This is how cities grow in the Netherlands, where a relatively high population density, a strong civic culture, and the peculiar difficulties of building on land below sea level, combine to favour strict public control.

At the other extreme are 'loose' planning systems that put no limits on urban spread and merely regulate the density and type of development as a safeguard against bad-neighbour activities which might lower property values in a vicinity. This is how cities grow in Canada, the United States and Australia: countries where land is plentiful and the ethos of private property rights strong.

In contrast, there are also ineffectual planning systems that erect a legal framework of regulation over the location and type of surburban development without sufficient powers of enforcement to ensure compliance. Much of the scattered building activity by the poor around cities in Third World countries is illegal in contravening the (sometimes hopelessly unrealistic) plans for orderly suburban expansion to a high architectural standard. The same is true of a surprisingly high proportion of industrial and housing developments in the ugly outskirts of southern European cities such as Athens, Lisbon and Madrid. These are the result, not of permissive, but of ineffectual planning systems without legal teeth. The remainder of this chapter takes a closer look at two planning systems that are both legally effective but which stand in contrasted positions along the policy spectrum between Dutch restraint and American profusion – Japan and Britain.

JAPAN AND BRITAIN

Japan and Britain (to start with resemblances) are both prosperous, highly urbanised, densely populated, island nations with market economies. From its earlier history of industrialisation, Britain inherited a mature system of cities which in recent decades has been shedding both population and employment to small towns and rural areas. In Japan's more recent and phenomenally rapid development as an industrial power, cities have trebled their populations since 1945 and more than quadrupled in size. In both countries, the free operation of market forces would tend to disperse urban development pressure at low densities over very wide areas. But these forces have been curbed by a planning system that defines a boundary between town and country. Britain introduced effective planning measures immediately after the Second World War, as a delayed policy reaction against the scattered development characteristic of the late 1920s and 1930s. Japan's planning system dates from 1968 and was introduced in an attempt to stabilise the rampant land price inflation associated with property speculation around the fast-growing metropolitan centres.

Planning mechanisms

Comparing urban plans from the two countries, we notice some immediate similarities. Both countries use large scale maps of the urban and adjacent area and define areas of land use and building densities permitted in this or that zone, the lines of roads and other major infrastructures, and the sites of proposed new developments. The crucial boundary beyond which urban developments should not occur, at least in the time period of the plan, is generally marked in British plans by a positive, protective zone such as the 'green belt'. There is also a category in Britain left unmarked as so-called 'white land'. This category also restricts development, since it is presumed that only land specifically allocated for building purposes should be so used.

In Japan the city plan classifies all land unambiguously as either Urbanisation Promotion Area (UPA) or Urbanisation Control Area (UCA). The UPA contains the already built-up area plus an additional capacity for a decade's further expansion. The UCA is, nominally, an area prohibited for development. The process of defining the UPA–UCA boundary – *senbiki* or line-drawing – is the political heart of planning in Japan, just as the designation of green belt is in Britain. Because of the sensitivity of this task, and the detailed local knowledge it requires, it is devolved in both countries to local government under the general supervision of a central ministry.

Land management

If the mechanisms of planning in the two countries are broadly comparable, the management approach could not be more different. British planning authorities operate what could be called a tight system of land release. Local authorities are required to designate sufficient land for a five year supply of housing, according to a procedure laid down by central government. The allocation is essentially conservative, being based upon past trends. It does not allow for any radical fluctuation in the growth rate of an urban region – indeed, the purpose of the system is precisely to dampen down such fluctuations. For example, throughout the 1980s, when the economy of south-east England was booming, the planning system continued to release land at the rather slow and steady rate of the previous two decades. This produced major constraints on industrial and housing development.

The thought uppermost in the minds of British planners, as they draw the line allocating the ration of land for development, is that they must minimise the impact of new building on the countryside. In fact, new building is almost entirely precluded on the edge of large and medium sized cities, where long established green belt policies have fossilised much of the urban boundary as it was in the mid-1950s – or, indeed, at the outbreak of the Second World War. Demand for new homes, shops and factories is diverted away from the fringe of major centres to locations selected by the planning system. Where possible, developers are encouraged to use up spare land by in-filling sites within existing urban areas in preference to green field sites. The recent 'deindustrialisation' of British cities has, of course, been a plentiful source of surplus inner city acreage. At present, over half the land used by the house building industry is obtained from the recycling of older urban sites. This high proportion is regarded by government as an indicator of the success of the planning system. It relieves pressure for development in the countryside, thus protecting the environment, and helps the regeneration of run-down sections of cities.

In contrast to the tight corset around British towns, the Japanese planning system is loose and commodious. It is formally required to allocate sufficient land for ten years, twice as long as the British equivalent. The extension of the planning boundary rests upon highly generous assumptions about the rate of urban development. In fact, the current stock of land available for development in Urbanisation Promotion Areas in Japan should suffice well into the twenty-first century. In addition, no attempt is made by Japanese city planners to use up existing land within the built-up area before rolling the urban boundary forward. Also, for reasons explained later in this chapter, the cities contain remarkably large amounts of undeveloped farmland – over 17,300 acres within Tokyo, for example, even though this metropolis has the world's highest property values and one of the densest populations.

Effects of planning

The effect of the different planning regimes is clearly visible in the physical form of cities in both countries. British cities, for the most part, are fully built-up to a cleanly defined edge reminiscent of the sharp boundaries of pre-modern settlements. This pattern is highly unnatural under modern conditions of urbanisation, and results from the careful management of the land resource by local planning authorities. Its typical result is shown in Plate 10.1.

Japanese cities have a looser texture, with paddy rice and orchards densely interspersed through built-up suburbs. Travelling outwards, we do not come to definite boundary as we have in Britain. Buildings remain scattered through the landscape, although their frequency diminishes with distance from the urban centres. The edge of the Urbanisational Promotion Area is not tangible, though close examination of Plate 10.2 will show a change in the texture of

MICHAEL HEBBERT

Plate 10.1 Drawing the line in England. In Barnet, 17 km from the centre of London, the inner boundary of the metropolitan green belt defined in the 1950s coincides exactly with the physical edge of the metropolis today. (*Photo*: Hunting Aerofilms)

development towards single-unit dwellings and small terraces, the so-called 'pocket handkerchief' developments. Another feature of the Urbanisation Control Area is factories, schools, hospitals and other large establishments located out amongst the paddy fields, as well as the lines of major new roads.

The age of cities accounts for some of these differences. Japanese cities are young by comparison with British ones. They have mushroomed in the last three decades and have the gap-toothed appearance of pioneer cities that need time to consolidate. Yet that is not the whole story. Older settlements in Japan also have the same open form, whereas newly built British suburbs tend towards an almost exaggerated neatness around their outer boundaries. The different approaches to the management of urban growth reflect deep-rooted contrasts in the relation between people and the land.

Plate 10.2   Drawing the line in Japan. A section of the boundary between 'promotion' and 'control' areas on the edge of Sayama city, a commuter suburb 35 km from the centre of Tokyo. About half the developments in the control area were built after 1969, when the line was drawn. Some of the larger developments are highlighted with arrows. (*Photo*: Japanese government, National Geographical Survey Institute)

The two key factors are, first, the pattern of land ownership, and second, the nature of farming.

Origins of planning attitudes

If we go back to the Middle Ages we find a recognisably similar pattern of land ownership in the two countries. Land was owned by a feudal aristocracy and worked by a mass of peasants on small plots. Thereafter the histories diverge sharply. In Britain the enclosure movement in the eighteenth and nineteenth centuries, and the modernisation of agriculture, drove the peasants off the land. The land was divided into large holdings, owned by a smaller number of individuals, and farmed with a minimum of labour. No

211

such process occurred in Japan, or indeed in any oriental country. The peasant population has remained tied to the land by the labour-intensive nature of rice cultivation. At the end of the Second World War the American occupying forces carried out a land reform which gave 10 million peasant farmers full ownership of their small plots. To this day, almost all land around fast-growing urban areas is owned by these small scale farmers, the majority of whom also have an income from an urban job in a factory or in the service sector. In suburban areas it is the small farmers who own the luxurious tiled-roof detached houses, surrounded by expensively transplanted forest trees. They are the new rich owners of the country's land – its scarcest and most precious resource.

Now consider the attitude towards urban growth in the two countries. Britain has an overwhelmingly urbanised population. Only one person in a hundred works on the land. For the rest of the population 'the countryside', as we call it, has become a potent symbol of escape from ordinary workaday life. Some sociologists argue that this strongly felt attachment to things rural results from the influence of a top-heavy property ownership structure, in which 1 per cent of the population owns 71 per cent of the land, and from the continued social status of the landed aristocracy. However, it is also true that urban development is popularly regarded in Britain as a negative process that 'spoils' something of intrinsic environmental value. As a result the concern of the planners to operate a tight system of land release, which minimises the green-field incursions of building activity, seems to have wide public support. The strict policy of urban containment used to be defended on grounds of an island nation's need to maintain self-sufficiency in food production. As the European Community now has chronic problems of agricultural surplus, the argument for so tightly rationing building land in Britain has had to shift to different and perhaps more honest ground; a cultural preference and an environmental concern for green landscape uncluttered by building.

The notion that planning should attempt to preserve 'countryside' is utterly alien to Japanese thinking. The dyked, irrigated and constantly tended landscape of paddy fields holds for the Japanese none of that romantic illusion of communion with nature which the British find in their depopulated green pastures and arable fields. Japanese farmers, who constitute 8 per cent of the population, are like peasants anywhere, unsentimental about the holdings to which they were historically tied in an unending cycle of manual toil.

Where possible (which means especially within travelling distance of the fast-growing centres of industrial and service employment), they seek the opportunity to realise some of the huge capital value of their land by selling it for building. The average size of holdings is small, typically two acres, and the intelligent owner will not bring all of it on to the market at any one time. The result can be seen (Plate 10.2) in the scatter of widely dispersed and very small developments. As owners release further slivers of land, so the area becomes more fully built-up. But deep in the inner suburbs of Tokyo and other great Japanese cities there remain abundant small pieces of open land – sometimes used for parking, but often preserved (for tax reasons) as paddy rice or vegetable plots. Kept unbuilt they are a store of wealth, accumulating value at a rate that has outperformed all other investments.

Farmers around all major urban centres in Japan lobby energetically to have the boundaries of Urbanisation Promotion Areas drawn as widely as possible, so that they may build as and when they wish. The planners, however, know that the thin spread of housing developments which accords with farmers' preference will cost the taxpayer more to service with sewers, schools and other infrastructure. In an effort to raise levels of urban service provision – which are extremely low in Japan compared to its high level of economic prosperity – they try to draw the line as tightly as possible. Unlike their British counterparts, they are not concerned with the physical containment of the built-up area as an end in itself, only with preventing a scatter of private building without facilities. If farmers in a control area can strike an agreement to combine their holdings into a properly serviced development scheme, their land will automatically be rescheduled as an Urbanisation Promotion Area. The planning system is there to protect not the countryside but the municipal budget. In fact, one of the most striking geographical consequences of the Japanese planner's intervention is the proliferation of public facilities (clinics, schools, administrative offices, pumping stations, refuse depots, and the like) in the control areas around Japanese towns, where land is half the price, even though these services are often inconveniently located from a functional point of view.

CONCLUSION

In Britain and Japan alike the regulation of urban growth has clearly had significant repercussions on the land and property markets. Its economic dimensions – like many of the geographical interventions discussed in this book – are complicated and important enough to have generated a good deal of political controversy, and an academic literature to match. The essentials are clear. Planning, by rationing development opportunity, has the effect of piling up value on certain sites at the expense of others. The tighter the line is drawn, the greater the difference planning makes.

In Britain, with its tight system, a shift of the line can add hundreds of thousand of pounds to the value of the land concerned. A one-acre paddock on the edge of a town within commuting distance of London might be worth say £10,000 in its present use, but acquires a value of £300,000 if designated as building land by the local council. The price difference to either side of the line widened rapidly in the 1980s. While average prices per acre of farmland in the south-east rose from £1,360 in 1980 to £1,860 in 1986, the building land average leapt upwards from £65,500 to £195,500 over the same period. Town planning's role in this is hotly disputed. The development industry argues that planning has directly caused the land price inflation (and so rising house prices) by over-restrictive policies for countryside protection. Planners reply that the physical supply of land has been carefully matched to demand. Its increasing value is a consequence, not a cause, of rising house prices, which in turn reflect rising disposable income as well as taxation and social policies that favour home ownership.

Comparative evidence supports the planners' contention that a looser system of urban land supply would not necessarily bring prices down. In the Japanese case the reverse is true: despite a consistently lavish allocation of land for development, both urban and agricultural land values are the highest in the world. Amongst the many contributory factors to these astronomical price levels the geographical management of urban growth plays a part. A loose-fit system offers the majority of landowners in a metropolitan region the prospect of bringing land on to the urban market. The general level of farmland prices rises accordingly. No longer linked, as in England, to the economics of agriculture, it reflects the money which well-to-do town dwellers are willing to spend on a small house plot. But perversely, from the consumer's point of view, landowners prefer to have the option to build rather than to exercise it. They use unbuilt land as a store of wealth: the less they release, the more demand builds up, and the more value it accumulates.

The present author's research has shown the painfully slow

construction rate within the generously designated Urbanisation
Promotion Areas of suburban municipalities around Tokyo. In the
new, fast growing cities on the commuter lines in the outer
metropolitan region, at least 10 per cent of the surface area is vacant
land, already serviced with roads but not yet available for building.
Throughout the extensive zone allocated by the planners for
urbanisation, farming and building coexist higgledy-piggledy. Over
a long period of time the Japanese government has contemplated tax
policies to discourage landowners from hoarding, to get land on to
the market, to ease supply and so to bring prices down. As in the
British case, the strictly geographical intervention of drawing the line
has proved to have some undesirable economic consequences that, in
turn, stimulate the need for appropriate taxation and other measures.

Both Britain and Japan are densely populated, industrialised
islands where the management of urban growth is recognised in the
late twentieth century as an essential task of government. Public
intervention has become a dominant force in the land and property
markets and in the social geography and physical appearance of
cities. Though its effect in Britain is strong enough to create an
illusion of continuity with the compact city of the pre-modern era,
'drawing the line' is a distinctly modern phenomenon.

A GREEN POSTSCRIPT

Environmental concerns have quite suddenly penetrated British
land use policy at both national and local levels since 1990. The
relationships between sustainability and the physical pattern of
development are complex and not fully researched as yet. Most
experts, however, are agreed on three urgent needs: to reduce
dependency on the private motor-car; to make power generation
more efficient (particularly through combined heat and power
schemes); and to promote more efficient domestic energy use.
Urban form is a significant factor in all three. Contained urban
settlements at moderate densities can be environmentally more
friendly than dispersed low-density urbanization. People may be
able to walk, cycle and use public transport instead of the polluting
motor-car. Local power stations may be able to recycle their waste
heat productively through district heating instead of releasing it into
the atmosphere. Domestic heat loss is usually lowest when housing
layout is compact. Urban compactness is a necessary though not a
sufficient condition for sustainable development, and in the 1990s

has become one of the principal motives for drawing a line between town and country.

QUESTIONS

1 What factors kept early towns and their urban regions so limited in extent?
2 What factors have allowed modern towns to spread over such great distances?
3 Why does land speculation sometimes lead to land being in a limbo of development?
4 List the main factors which lead to different geographical patterns at the fringes of British and Japanese cities.
5 Compare the 'tightness' of the planning systems in Britain and Japan and examine how they affect land prices.

FURTHER READING

Best, R. (1981) *Land Use and Living Space* (Methuen: London). Though slightly out of date this remains the most straightforward review of the facts and figures about town and country use of land in Britain and other countries.
Jones, E. (1990) *Metropolis: The World's Great Cities* (OPUS: Oxford). The later chapters of this wide-ranging text contain many stimulating ideas about the great cities' appetite for land, now and in the future.
Mather, A. S. (1986) *Land Use* (Longman: Harlow). Highly recommended both as an introduction to the geographical analysis of land use and as a text on planning and environmental policy.
Local planning authority *Statutory Plans*. Under the Town and Country Planning Acts all local councils in Britain prepare plans for their town and country areas which can be inspected at the offices of the planning department.
McDonald, D. (1985) *A Geography of Modern Japan* (Paul Norbury: Ashford). A good general introduction to the geographical setting of modern urban development in Japan.
HM Government (1990) *This Common Inheritance* (HMSO: London). Essentially classroom support material on the debate on sustainable development in Britain.
Blowers, A. (ed.) (1993) *Planning for Sustainable Development* (Earthscan: London).

11

MANAGING URBAN CHANGE
The case of the British inner city
Derek R. Diamond

THE INNER CITY

Inner cities have proved to be a topic of continuing concern in Britain and most other developed countries. In Britain the need to give inner cities special support was specifically recognised by the government in an inner city White Paper of June 1977 (Cmnd 6845). The period since has seen, in Britain, the other European countries, the United States and elsewhere, very considerable government involvement in a variety of programmes (often involving the creation of new agencies) designed to help tackle the special problems of these areas. However, the story of government involvement in inner cities begins much earlier and is likely to persist well into the future. Essentially this is because as great social and economic changes have occurred cities have had to adapt by expansion, contraction and redevelopment – processes that can bring considerable disruption and hardship to sections of their populations.

Inner city problems can be best understood as issues resulting from the underlying forces, outlined in earlier chapters, that have been changing the economy and social structure of urban areas. Some of these forces are international, some national, some local; but they all interact with the specific conditions existing in large cities to produce a variety of outcomes, now conventionally labelled 'the inner city problem'. Peter Hall was surely right when he wrote in the aftermath of the city riots of 1986 (see Figure 11.1):

> We are again obsessed by an inner city crisis. But the real problem is not an urban one at all: it is a bundle of economic and social problems that happen to impact with especial force

Figure 11.1 Poverty and urban riots, Greater London

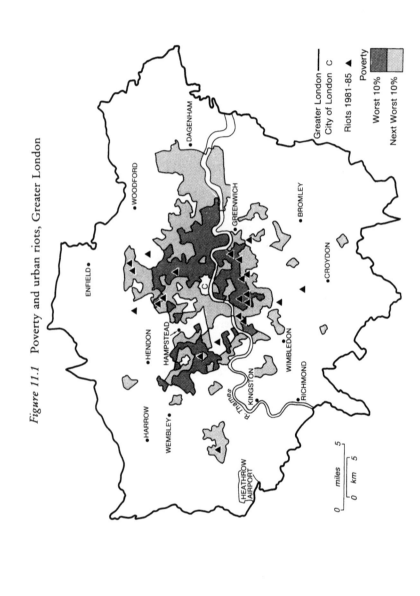

Greater London ⎯
City of London C
Riots 1981-85 ▲

Poverty
Worst 10%
Next Worst 10%

DAGENHAM
WOODFORD
GREENWICH
BROMLEY
ENFIELD
CROYDON
HENDON
HAMPSTEAD
WIMBLEDON
HARROW
WEMBLEY
KINGSTON
RICHMOND
Thames
R. Thames
HEATHROW AIRPORT

0   miles   5
0   km   5

on people in some parts of the cities. These problems are very
deep and intractable; they have their roots in the development
of our economies and our societies – not just here in Britain,
but also in the other Western countries – over the last thirty
years.

Urban historians have pointed out the similarity of today's inner
city problems to those in the second-half of the nineteenth century.
The paradox is that, despite its considerable history, the agreed
identification of its problems and its presence in many countries,
there is no single widely accepted definition of what an inner city
is.

Geographical analysis of Western cities has consistently revealed
what has been called a 'zone of transition'. This is the part of the
city most seriously affected by Hall's 'bundle of economic and
social problems'. It is typically composed of mixed commercial,
residential and other land uses, with considerable vacant land and
poorly maintained properties. It separates the commercial and retail
heart of the city from the surrounding suburban residential or
industrial districts.

The inner city as discussed in this chapter corresponds to this
transition zone: it is an area that surrounds, but excludes, the city
centre (or commercial 'downtown') and is itself surrounded by, but
excludes, the outer suburbs, which are usually of twentieth century
age and have a much more segregated land use. In Britain the zone
of transition was first developed for urban purposes in the nineteenth
century and, since 1945, has been the focus of several major public
policy and planning initiatives. In a sense this zone is made up of
what are now the 'inner suburbs', with intensive development and
high population density.

The severity of the problems encountered varies between cities
and between countries. It is the context of the inner city rather than
its location, *per se*, that explains what is happening to it. The major
social aspect of this phenomenon has been the concentration in the
inner city of what is frequently termed an 'underclass' population.
These are the poor and the disadvantaged who find it difficult to
participate in the changing urban economy. Often this reflects their
recent arrival, as in the case of blacks and Asians in Britain, or the
foreign workers and East European migrants in continental Europe
(particularly in Germany). This is well illustrated by the following
extract from a report on Berlin in 1978.

The symptoms of the 'Kreuzberger Illness' are not new and are not unique to Berlin. They are repeated in many other cities: the physical condition of the inner city area has deteriorated, because maintenance has been neglected and modernization never begun, and the older inhabitants who can afford to move away, leaving room for foreign workers. Thus, finally only those that must have the cheapest accommodation stay in the area; the elderly, foreign workers, and those who must live at the fringe of the welfare state. This 'worsening' of the residential structure has its effects in the local economy, and on the willingness of property owners to invest. The area loses its economic viability, and deteriorates further.

This chapter first describes the key features of contemporary inner city problems and then seeks to explain the nature and impact of the public policy response.

FEATURES OF THE INNER CITY

As recently as 1960 it was still possible to find in many inner cities the actual infrastructure put in place in the early stages of the industrial revolution. This would have been true in cities as varied as London, Glasgow, Manchester or Birmingham. This infrastructure was characterised by multi-storey factories built for steam power, and located beside canals and railways. The factories were surrounded by dense terraced housing which had, with the rare exceptions of 1930s slum clearance schemes or 1940s blitz damage, survived almost totally intact. In 1960 Belfast's inner city was described in these terms:

> Against a setting of mills and smoke stacks and gantries, are set the endless rows of countless small, identical, workers' houses which the last century has bequeathed to this. On the whole conditions are mean, cramped and crowded. Every corner has its shop, every vista has its factory, and every alleyway is a playground for scores of children.

Similar descriptions would have characterised most northern industrial cities in Britain, the East End of London, and cities in the Ruhr, north-eastern France and the US manufacturing belt.

The built environment of the great urban expansion of the

nineteenth century which created the large industrial city in northern
Europe and in the industrial belt of the USA had, not surprisingly,
by 1960 become seriously outmoded. Obsolescence took many
forms. Housing was sub-standard in terms of its size and facilities,
with many houses lacking indoor toilets or bathrooms; factories
had poor road access and unsuitable premises for new manufacturing technologies; almost all canals and many railways and docks
were virtually unused; the public infrastructure of hospitals, schools,
sewers and water supply was in need of massive repair and
maintenance, and the road pattern was unsuited to, and inadequate
for, the rising tide of motor traffic.

In contrast, the twentieth century expansion of the industrial city
took a different form. Residential suburbs of modest density
were dominated by detached and semi-detached houses with minimum building standards imposed by successive governments. Such
suburbs were often separated from the new industrial estates; they
had single storey plants and planned road access, via purpose-built
roads and dual carriageways. Let us examine the various dimensions
of these changes.

Population change

Major changes in the cities were reflected, not surprisingly, in
population geography. The numbers living in the inner cities
reached a peak around 1930 and subsequently diminished, without a
break, until the present. Large numbers left for the outer suburbs
and beyond. In the 1960s and 1970s this migration included those
moving to new towns, expanded towns and overspill sites. At the
same time large flows of new residents entered the inner city,
particularly immigrants from the Commonwealth countries. Thus,
although overall population declined, a radical change occurred in
its composition. The distinctive age structure, household composition and ethnic characteristics of present inner city populations are
the outcome of complex trends that have operated over several
decades.

The out-migrants have largely been families of young adults with
children seeking a better environment. The suburbs offered housing
at lower residential densities, houses with modern facilities, new
school buildings, pedestrian precinct shopping centres and high
levels of public facilities such as swimming pools, parks and better
road provision. The demographic bias is equally true of those who

moved to owner-occupied homes as of those who moved to peripheral estates of council housing, where families had priority in the allocation scheme. Those who found it difficult to join the 'suburbanisation' process were the poor who could not afford the increased costs of an improved environment, and those who had recently arrived in the inner city and were therefore low on council house waiting lists. For such people, with few skills, an inner area labour market seemed to offer reasonable job prospects in unskilled, if low paid, work. The result, therefore, was a process of 'social filtering'. This gave the inner city fewer family households but large numbers of single person households (including lone pensioners and single young adults), as well as a higher proportion of single parent families, the unskilled and the unemployed. This created the conditions for the emergence of an 'underclass'.

In addition, the inner cities proved especially attractive in the 1950s and 1960s to coloured immigrants who settled, usually in close proximity to each other, in the inner areas of many of the large industrial cities. Of crucial importance to the emergence of these ethnic concentrations was the place and nature of their employment. Early post-war immigrants were recruited to work in public services such as local transport and refuse collection, and as non-medical hospital personnel, all of which was low skill, low paid employment located predominantly in the inner city. It was precisely in such locations that cheap housing was available in the private rented sector. Accommodation was relatively inexpensive because it was in poor condition, often being 100 to 150 years old and lacking in modern amenities. Without the resources to buy their own houses, and without sufficiently long residence to obtain a council house, immigrants had little choice. Further, many subsequent immigrants were relatives or friends of those already arrived, and it was to be expected that they would attempt to find accommodation in the same localities, thus reinforcing the concentration. In Britain the ethnic minorities live in varying degrees of mixture with the white population and not in conditions of total segregation as found with poor negroes in the ghettoes of American cities. But while the barriers of low income and recent arrival lessen over time, and West Indians in particular have become well represented in council housing, there is still a strong correlation between poor housing and ethnic status in British inner cities.

Economic change

It was the growing shortage of jobs in the inner city and the consequent rising level of local unemployment that led in 1970 to the introduction of new laws to restrict overseas migration to Britain. The economic decline of the inner city has been of long standing, but before about 1970 it was somewhat masked by overall national economic growth. However, in most older industrial cities it is clear that a steep decline in inner city employment began as far back as the 1960s, as a result of the level of manufacturing plant closures greatly exceeding the level of plant openings.

Some industrial premises closed because their products were no longer in demand (e.g. steam locomotives in Glasgow), others because their market was taken by more competitive producers at home and abroad (e.g. inner city ironworks and shipyards, etc.) Yet others, usually small workshops and the like, were driven out of business by slum clearance schemes. When local authorities involved in slum clearance offered new accommodation to 'back court' industries, only a small proportion of those displaced were able or willing to relocate within the inner city. The difficulty of adopting new production techniques was also acute for many inner city firms whose sites were too small and physically hemmed in by other buildings. Expanding businesses thus often moved to the outer suburbs, or beyond, while a spate of amalgamations and take-overs after 1965 provided yet another opportunity to improve industrial productivity by relocating production on more efficient sites.

Another important contribution to the economic decline of inner cities was the run-down of traditional services and activities such as docks, railways, distribution and warehousing. This was caused by the increasing accessibility of the urban periphery as the national motorway network came into being and changes took place in the technology of transport and distribution. The obvious effect of this industrial restructuring was not only the emergence of derelict industrial sites which there was little or no demand to redevelop, but also a decline in jobs, particularly for male manual workers.

This fall in inner city employment (a loss of 45 per cent of jobs 1951–81) has been greater than the fall in the number of the population of working age (a reduction of 35 per cent 1951–81). Thus unemployment in the inner cities, which even in 1951 was 33 per cent above the national average, rose to 51 per cent above the average by 1981. In many inner city localities unemployment rates became extremely high. In May 1986 the twenty-five worst wards in London averaged 25 per cent unemployment. But the rise in unemployment did not affect all inner city residents equally. Those who were black, disabled, semi-skilled or unskilled, elderly, without work experience (such as school-leavers), with no qualifications or from single-parent families were most affected. For those groups combining several of these characteristics the level of unemployment rose as high as 75 per cent, and these very high unemployment levels tended also to be concentrated in very small areas, such as poor private rental housing (e.g. in Brixton) and in many council housing estates.

The combination of population decline, changes in manufacturing production methods among the surviving inner city industries and the major shift of city employment from manufacturing to commerce has created a complicated mismatch between the residents and the skill requirements of the new local activities. Retraining to adapt local populations to new employment needs has proved surprisingly difficult, although major new efforts are now under way to tackle this 'skills gap'. The result is that levels of commuting out of the inner city to jobs elsewhere, including the city centres, is low, whilst conversely there has been a steady rise since 1951 in the share of inner city employment taken by commuters (some of them previously inner city residents) from outside the inner city. In 1981 it was estimated to be 39 per cent nationally, but much higher in London and Liverpool. As a result, the emergence of an underclass of the disadvantaged who remained outside the economic system was fostered in the inner areas of the large industrial cities.

Change in environmental living conditions

Although the government's 1977 White Paper pointed out that 'the single most characteristic feature of the inner city is the age of its housing' it is this aspect of the built environment that has altered most in the last forty-five years. By the mid-1980s most of the terrible housing bequeathed by the Victorians had either been demolished or improved. By 1981 96 per cent of national households enjoyed the exclusive use of a bath and inside toilet. In the core cities of the conurbations, however, conditions remained markedly less satisfactory, with nearly twice the national average proportion of households not having exclusive use of the basic amenities. For London the figure was 8·8 per cent, for Manchester 7·5 per cent and for Liverpool 7·3 per cent.

A truly enormous slum clearance effort (see below) was responsible for the dramatic improvement in national housing standards, but in the big cities the processes of clearance and redevelopment got badly out of step. As a consequence there is now a great deal of vacant land in many inner areas, much of it in public ownership. There is also much under-used land and property, with shops, old factories, and even post-war houses boarded up. A recent study of one of the largest urban renewal projects in Britain, the East End of Glasgow, concluded that the area now contains less than one-third of the numbers who lived there a generation ago. It has experienced some of the most disastrous industrial closures and some of the biggest slum clearance projects in the world, and most of its people have fled or been rehoused elsewhere. The jagged population pyramid looks much like that of Poland or Germany after the last war. What this community has passed through has been described as in some ways rather like a major war.

Between 1960 and 1968, in an attempt to retain inner urban populations and to utilise new industrial building techniques, just over 20 per cent of new public housing approved for central government subsidy was in the form of multi-storey flats. This form of redevelopment was, however, confined almost entirely to the inner areas of London, Liverpool and Glasgow and other major cities where land shortages were acute enough to outweigh doubts about the cost and possible social consequences. Although only 400,000 homes out of a total stock of British housing of 20 million (i.e. 2 per cent) were in tall flats, the poor and dense design of much post-war council provision, combined with unfinished neighbourhood plans, faulty local authority housing management and the growth of social problems such as drugs, vandalism and truancy, has created areas of great social stress, physical dereliction and drab environment. In such localities a sense of collective deprivation arises

From a pervasive sense of decay and neglect which affects the whole area, through the decline in community spirit, through an often low standard of neighbourhood facilities, and through greater exposure to crime and vandalism, which is a real form of deprivation, above all to old people.

(White Paper, 1977)

Deprivation

Although there are pockets of prosperity and high quality environment in inner cities, the declining flow of public resources in the 1970s and increasing unemployment drew attention to the major architectural and planning errors of the 1960s. This led to the recognition of 'urban priority' areas for which the Archbishop of Canterbury, among others, called for priority of attention. These were inner city districts of specially disadvantaged character suffering from economic decline, physical decay and social disintegration. These processes interlock and combine to create multiple deprivation such that the inhabitants of priority areas are 'prevented from entering fully into the mainstream of the normal life of the nation'. They constitute what has been termed the 'underclass', as noted earlier. In his call for action (*Faith in the City*, 1985) the archbishop noted the parallels between the 1980s and the 1880s in terms of increasing inequality and social disintegration.

Deprivation is extremely difficult to define but clearly implies that the quality of life and access to material goods of affected households fall below some generally agreed minimum standard. It is more than simple financial poverty, since it attempts to include welfare support derived from such local services as health and education. In England, in 1981, eight indicators were used by the Department of the Environment to identify the most deprived local authorities – they were:

1 The percentage of economically active residents who were unemployed.
2 The percentage of private households living at a density of more than one person per room.
3 The percentage of private households which contained at least one single parent family with dependent children up to fifteen years of age.
4 The percentage of private households containing a lone pensioner.
5 The percentage of private households which lack exclusive use of a bath and inside toilet.
6 The percentage of residents in households where the head was born in the New Commonwealth or Pakiskan (i.e. non-white).
7 The percentage change in the population 1971–81.
8 The ratio of the standardised local death rate to the national rate.

These indicators have been used to define cities with Urban Priority Areas (UPAs). Although these vary considerably in size, there is a major concentration in the large conurbations, as shown in Figure 11.2. The UPAs have now become the focus of a variety of public sector targeted support.

The problems of these areas can be understood from the case of Glasgow. In the 1970s the Census of Population in Scotland was analysed to reveal a clear spatial concentration in Glasgow of the distribution of multiply deprived people, as shown in Table 11.1.

Table 11.1 Deprivation in Glasgow, 1971

Measure	%
1 Proportion of all Scottish households	18·9
2 Proportion of all Scottish households with four or more deprivation indicators	31·3
3 Proportion of all Scottish households with five or more deprivation indicators	40·6
4 Proportion of the 10 per cent worst enumeration districts[1] in Scotland	51·0
5 Proportion of the 10 per cent worst enumeration districts in Great Britain	59·2

Note: [1] A small spatial unit used for collecting census data.

A similar study of eighty-five urban areas in Britain undertaken ten years later used a broader range of 1981 census variables. It found that in the list of the twenty-seven most deprived areas in Britain the Glasgow inner area was still the most disadvantaged. It was concluded that 'no other area in the whole country comes anywhere near Glasgow on the educational deficit. This alone would seem to ensure the perpetuation of the present position of half of Glasgow's population in the next decade.'

The persistence of such levels of deprivation has been associated, in some instances, with outbreaks of rioting in the inner cities. In the USA, in the late 1960s, riots occurred in such cities as Newark, Detroit and Los Angeles (Watts). In the UK, in the early 1980s, Bristol (St Paul's), Birmingham (Handsworth), Liverpool (Toxteth), London (Brixton) and Manchester (Moss Side) were similarly affected. (See Figure 11.1 for the case of London.) Such riots were unexpected and led to major public inquiries which came to remarkably similar conclusions. American studies found that the typical rioter was a teenage or young adult black, a long time resident of the city in

Figure 11.2 Urban Priority Areas, Great Britain, 1990. *UDC* Urban Develpment Corporation, *UPA* Urban Priority Area, *EZ* Enterprise Zone. 'Valley Initiatives' apply only to South Wales

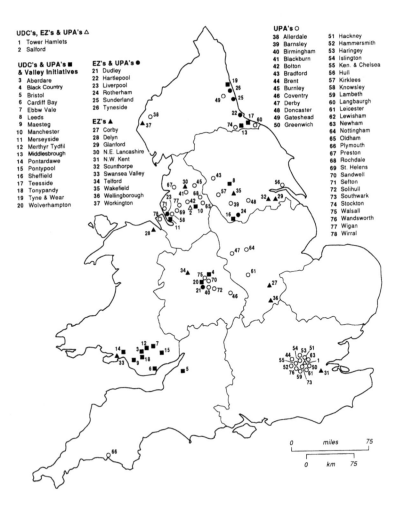

228

which the riot occurred, a secondary school drop-out (but often better educated than a non-rioting black neighbour) and usually unemployed or working in a menial job. The most fiercely held grievances were about police practices, unemployment and housing conditions. In the UK the picture was much the same. Thus the recommendations of the Scarman Report on the 1981 Brixton disorders followed quite closely those of the earlier Kerner Report in the USA. Both stressed the need for policies to end the multiple disadvantages that blacks and other groups experience, and the need to encourage the expansion of a black middle class. Lord Scarman put particular emphasis on education and training facilities for inner city residents in an attempt to bring the growing underclass into the mainstream of society.

This examination of how the population, employment and environment of the inner city have changed in recent decades, and of how these changes have interacted to affect the quality of life of many inner city residents, shows clearly why inner cities have continued to be the major focus of urban problems. This long history of economic difficulty and social stress led to a very considerable public policy response in the second half of the 1980s. It is to this story of public policy that we now turn.

THE POLICY RESPONSE

The previous discussion has indicated the considerable variety of significant changes – economic, social, demographic and environmental – that have been occurring in the inner city in recent decades. They have been mirrored in a bewildering series of government programmes, particularly since the 1960s, so that it is often extremely difficult to identify the specific results of this display of government concern. The broad pattern of policies is remarkably similar in both the UK and USA, despite their radically different local government and fiscal machinery. Policy shifts appear to reflect both the changing circumstances of the inner city and the currently favoured explanations of such changes. In broad terms, however, it is possible to identify three distinct phases in inner city policy – the *physical approach* of the 1950s and 1960s, the *social approach* of the 1970s and the *enterprise approach* of the 1980s. Although each of these phases is characterised by a different emphasis, it is also true that the policies of the earlier phases were normally continued in a lower key so that policy has become, with time, more comprehensive as well as more complex.

The physical approach

Physical development was the theme of urban renewal in the period after the Second World War. Not only was wartime bomb damage a stimulus to redevelopment, but also a change in social attitudes occurred in favour of a major improvement in the quality of housing overall. This was reflected during the Second World War by the development of propaganda for 'homes fit for heroes'. In any case, many existing inner city houses were surrounded by elderly industrial premises, inadequate roads and little public open space. Thus the need to rearrange the land use and street pattern to suit twentieth century requirements was clear. This was to be done by, for example, providing schools with adequate adjacent playing fields, pedestrian shopping centres and a car park space at home for everyone.

In order to achieve such objectives the density of population had to be massively reduced – by between 30 and 60 per cent in most redevelopment areas. There was thus a considerable displacement of population to peripheral housing estates, to new towns and to more distant towns embraced in expansion schemes under 'overspill' arrangements made by the local planning authorities. This massive urban redevelopment programme was administered by the local authorities, which, under the terms of the 1947 Town and Country Planning Act, could designate Comprehensive Development Areas where 'bad layout and obsolete development' existed. Coupled with the redevelopment of parts of the commercial centres of the industrial cities, this policy had the effect of encouraging decentralisation from the congested inner areas of the industrial conurbations so as to develop new towns and invigorate the smaller expanding towns. This simultaneously permitted redevelopment of the inner city at the lower densities suitable for a society with rising car ownership and rising standards of housing.

Inevitably there was massive physical disruption and also, it gradually emerged, immense social stress. This was not helped by the bureaucratic approach of the local authorities. The problems were then made worse by national economic difficulties of the mid-1960s which resulted in increasing uncertainty and delays in the programme. Consequently a new and less disruptive form of urban renewal was initiated at the end of the 1960s. Rather than redevelopment, rehabilitation, designed to provide a more acceptable built environment and yet retain social cohesion, became the practice in most cities. This was associated at first with the designation of General Improvement Areas (GIAs) and, in 1974, with a more socially focused version, the Housing Action Area (HAA). This approach involved special levels of improvement grant to home owners and tenants or landlords for housing improvements in small defined areas of older housing with at least a fifteen year life. The aim was to improve the whole locality markedly in a short period of time. Although such spatial targeting policies have been criticised because they result in public funds helping some who are not in need, there is clear evidence that public authority action was more effectively co-ordinated within the designated areas (GIAs and HAAs), and that this was to the benefit of all the residents. Also it

Plate 11.1   Urban regeneration in Dundee. Modernisation and refurbishment of blocks of flats (left) to introduce pitched roofs and individual gardens (right) on the Whitfield estate (*Photo*: Robert Bennett)

231

has been shown that private investment has been more readily attracted to old houses in adjacent areas.

The social approach

The shift in policy from redevelopment to rehabilitation, which brought a rapid end to the building of multi-storey flats, was in fact part of a major shift in perceptions of the inner city. In the United States, by contrast, there had not been the bomb damage from the war but there had nevertheless been large scale land clearance and comprehensive redevelopment. But by the mid-1960s, American policy had also shifted away from a 'federal bulldozer' strategy to a series of programmes designed not only to improve the physical fabric, but also to assist the residents of the inner city to overcome some of their difficulties. Under such labels as 'model cities', and 'economic opportunity' the federal government attempted to link public spending on urban renewal and environmental rehabilitation with the provision of social services, and in particular with the provision of low and middle income housing.

Although in retrospect most of these American programmes have been found to be rather unsuccessful, at the time the broader basis of the programmes seemed relevant to the emerging UK situation. This was recognised in a series of official reports: on London's housing (1965), primary schooling (1967) and the co-ordination of social services (1968). The result was a series of area focuses for policy. Thus 1968 saw the introduction of the British Educational Priority Areas, following the example of the American Headstart programme. In 1969 the British Urban Programme and Community Development Project package was modelled on the US Community Action Program, while the early 1980s saw Urban Development Grants, which were similar to the US Urban Development Action Grant Program. The widening basis of government policy can also be seen in the growing involvement of the Department of Education and Science and the Home Office as well as the Department of the Environment.

One argument more than any other formed the basis of these programmes. This was the thesis that deprivation was transmitted through a culture of poverty. This thesis held that deprivation can be transmitted from one generation to the next, so that the underclass would be self-perpetuating. Many children born to poor parents start school at a disadvantage. They receive little support from their parents and often find themselves in poorer schools with an above average turnover of teachers, which reduces their interest in attending. With a record of truancy and with few, if any, educational qualifications, they have difficulty in obtaining work and a decent wage. The poor quality of housing also contributes to below average health, adding to their difficulties in finding and keeping employment. There is therefore a high probability that this inheritance of low income, poor housing, inadequate education, low skill, poor health and propensity to crime will be passed on again to the next generation. These problems appear to be heightened in many single parent families.

It is important to recognise that this explanation suggests that intervention by public policy is likely to be unsuccessful if focused on a single point in the 'cycle of poverty'. Policy will turn the problem round only if several inverventions are concurrent. Action under the Urban Programme has therefore taken many different forms, but in its first four years (1969–73) most effort was spent on day nurseries, nursery education and child care in order to focus on children, especially of single parents.

These socially oriented inner city initiatives, and most of those that followed in the 1970s, defined areas of need, not in terms of physical decay, but in terms of economic and social character-istics such as the degree of housing stress, the number of single parent families and lone pensioner households, the proportion of Commonwealth immigrants, and indices of youth unemployment. Complex social indicators were devised to indicate the most deprived localities in order to ensure that government aid was targeted to those most in need.

One new feature of these programmes was the attempt to work with local community organisations in the Community Develop-ment Projects (1969–76). But this experiment fared no better than the American experience of working with similar grass roots organisations. Problems arose from the conflict that such pro-grammes engendered between central and local government and from the wide variety of objectives that community organisations followed, which frequently sought to maintain, rather than change, local society.

However, the Community Development Projects did show that the accelerating pace of economic change was creating further severe problems for inner city residents. This was confirmed by several government inner area studies undertaken between 1972 and 1977.

Inner Liverpool, Birmingham (Small Heath) and London (Lambeth) were studied in an attempt to find a 'total approach' to understanding the inner area problem. In their conclusions these studies emphasised the importance of the flight of capital and skilled labour from the inner city. Their key recommendation was thus the need for action to counter economic as well as environmental and social problems. This was taken up in the 1977 White Paper *Policy for the Inner Cities.* The government argued that:

> There is undoubtedly a need to tackle the problems of urban deprivation wherever they occur. But there must be a particular emphasis on the inner areas of some of the big cities because of the scale and intensity of their problems and the rapidity of run-down in population and employment.

The White Paper's conclusion that 'The decline in the economic fortunes of the inner areas often lies at the heart of the problem' was thus reflected in a further shift in policy.

The enterprise approach

Thus even before the election of the first Thatcher government in 1979 the changing perception of the problem had begun to alter the nature of policy. Though area-based 'positive discrimination' measures continued to have an orientation primarily to social services, the new emphasis was strongly on economic regeneration. It was recognised that one of the major stimuli to getting social development in inner cities was to stimulate new jobs to bring the socially disadvantaged into employment and thus participate in the material benefits of society. The existing regional industrial policies, which had dealt chiefly with larger areas like the north-east, South Wales and Merseyside were modified to accommodate an inner city emphasis. Metropolitan authorities were encouraged to reorientate their mainstream policies towards assisting industry. The Urban Programme was expanded to include upgrading the city environment, and local authorities were given greater powers to assist industry. There was a shift of financial resources to metropolitan areas through the annual central government support to local authorities (the Rate Support Grant) to provide a more unified central and local government response.

Since 1979 the Conservative government has retained the Partnership Areas and Programme Authorities established by the 1978

Inner Areas Act. But it has restricted local authority expenditure and encouraged private sector involvement, in order to emphasise what it termed 'wealth-creating activities'. However, there was a considerable shortfall of funds owing to the inability of the inner cities to meet the requirements of the government's urban initiatives. Thus, consistent with the government's aim of fostering private-led economic revival, two new measures were enacted both of which have drastically curtailed the local authority role.

First, in 1980, 'Enterprise Zones' were introduced. These were to be inner city areas, not exceeding 600 acres in extent, in which local rates and central taxes on business were reduced. In addition to financial incentives for new enterprises, less regulation and reduced state intervention applied in the form of simplified town planning procedures, exemption from industrial training levies and Development Land Tax, and fewer official requests for statistical information. These essentially property-related measures aimed to encourage the reclamation of derelict land and buildings. Major examples of EZs are in the Isle of Dogs in London, Trafford Park in Manchester, Belfast in Northern Ireland, and Clydebank in Scotland. The improved supply of property was expected to increase activity in manufacturing and services and thus lead to the creation of employment. In their first two years the original eleven enterprise zones attracted 725 firms and 8,000 jobs, but outside the London docklands EZ research has shown that in many cases the majority of the firms involved were short distance movers seeking the EZ subsidies. Only a few new or radically expanded enterprises were established. Although there are now twenty-two enterprise zones, their impact will remain small and local, particularly as the abolition of Development Land Tax and simplified planning procedures now apply to the whole country.

The establishment of urban development corporations (UDCs) in the London and Liverpool docklands in 1981, and in eleven other places in 1987, was the second major urban initiative of the Conservative government. These bodies, modelled on the development corporations that built the post-war new towns, take over local planning and have special powers and resources to promote development by land preparation, infrastructure provision and making loans and grants for building work. In the London docklands considerable commercial development for the financial services sector is under construction. For example, the Canary Wharf project alone will have a capacity for providing 50,000 new jobs. Also, several major national newspaper printing plants have moved from Fleet Street, while residential development for owner occupation is rapidly expanding, and the average price of land has increased almost tenfold. This early economic success tends to contrast with other areas which do not have the advantage of proximity to the City of London. However, the eleven UDCs created since 1989 are now also showing considerable success in physical redevelopment.

Plate 11.2   Regeneration in London's dockland. Office development and a flyover for the new light railway on the Isle of Dogs (*Photo*: Robert Bennett)

The emphasis on involving the private sector in urban renewal, apparent in the terms of reference for the UDCs, was again obvious in the revival of the Derelict Land Grant and the establishment of land registers listing all vacant or under-used land in public owner-ship. It was most obvious, however, in the introduction of the City Grant (formerly known as Urban Development Grant 1982, and its successor, Urban Renewal Grant 1987). This is designed to lever private investment into the inner cities by offering government financial support to privately funded development. The concept was initially closely based on the Urban Development Action Grant (UDAG) programme begun in 1977 in the USA as part of President Carter's attempt to create a national urban policy.

The UDAG programme was described as combining three innovative features: reliance on private investment; government grants based on tightly defined criteria to select from among competing applications; no funds released until the private capital is actually committed. In the UK by 1987 over 200 projects involving some £130 million of government funds had attracted an additional £520 million of private sector funds, thereby generating an overall leverage ratio of 1:4. But, again, failure to take up all the government funds available has shown how difficult it is to turn round the low demand for land and property in the inner city. Rapid growth in such demand between 1987 and 1989 led, for the first time, to major expansion of demand for City Grants, but the slowdown in the property market in the early 1990s has again made it difficult to attract private funds.

The government has also set up further agencies – broadly based City Action Teams and Task Forces. These are best described as co-ordinating bodies set up by central government departments. In addition there are Housing Action Trusts, Inner City Compacts for Schools, Education–Business partnerships, grants to local enterprise agencies and a stream of other specifically focused programmes to assist and encourage economic activity in the so-called 'Target Areas' – sixty-five in Scotland and seventy-three local authorities in England and Wales. These focus on the inner city UPAs, but also now include many deprived localities outside the inner areas of the large industrial cities. Most important of these deprived areas are the 'peripheral' housing estates, particularly in Scotland, where many former inner city residents were moved in the 1960s and 1970s during comprehensive redevelopment programmes.

The increasing spatial coverage and range of government initiatives were described in a government report, *Action for Cities*, launched by the Prime Minister in 1988. This report is remarkable in not containing a single new policy but seeking to draw numerous central departments into a coherent strategy. The strategy covers special arrangements for housing improvement, school education, skill training, employment generation and environmental improvement, and involves central and local government, private enterprise and the voluntary sector in many different agency agreements.

Given the many existing avenues of support, it is perhaps not surprising that new urban policies have not been developed since 1988. Some observers, however, remain critical. In August 1989 an Audit Commission report noted the lack of Whitehall co-ordination, the excessive red tape, the involvement of too many ministers and Treasury intransigence. It recommended urgent consideration of 'a single one-stop unit which can guide private sector investors through the maze of regulations which exist, and identify for them the relevant public, private and voluntary bodies needed to assist'. Indeed, although *Action for Cities* estimated that about 3,000 million would go into urban regeneration in 1988/89, bringing in several times that amount in private investment, there are of course other people-based government programmes (such as social security, state pensions for the retired, as well as education and training programmes) that also impact on deprived areas and, particularly, the inner city. Nor is this the total picture. Local authorities are involved on a regular day-to-day basis in coping with inner city change, as, for example, in closing schools as enrolment declines, strengthening the police presence in the face of increasing fear of crime, reallocating vacant local authority houses, promoting the area for business purposes and a thousand similar bureaucratic actions which, in aggregate, influence both the nature of the problems and the effectiveness of the remedial policies.

CONCLUSION

There can be no easy answer to the question of how far inner city areas have been changed by this large, lengthy and varied bundle of programmes. Clearly there would have been substantial changes in the inner city as a consequence of the impact of the powerful economic and social structural trends mentioned in the first part of this chapter. Almost every analyst believes that in the absence of policy the situation would have been worse. How effective the policy initiatives are judged to be, however, depends very largely on the test that is applied. Some would suggest the reduction of poverty and an end to the powerlessness of the poor are appropriate criteria. Others argue that the provision of a minimum level of shelter and income, combined with genuine opportunities for social advancement, are a better test. But the evidence of many local evaluations is that there has been a bewildering variety of outcomes: some places have 'succeeded' and others have 'failed' in their attempts to adapt to change.

There also remain questions about how to intervene effectively to use lessons learnt from the past. Should a regeneration strategy be predominantly place-oriented or predominantly people-oriented? Should the major objective be to raise the local authorities' capacity and resources to enable them to tackle their own problem areas, or should there be purpose-built, fixed-life agencies funded and operated directly by the central government, or some kind of partnership scheme? Whichever intervention style is adopted, how is the complex matter of appropriate collaboration between all four major parties – citizens, entrepreneurs, local and central government – to be accomplished? Some of the main failures identified are:

1 Too few net new jobs have been created from all the activity.
2 Very few of the jobs which have been created have gone to the most deprived because of (a) the leakage of benefits through open labour markets; and (b) lack of skills of the inner city population.
3 Where success has been achieved it has invariably been on the back of considerable sums of public expenditure, so that the supply of public spending is the control of the rate of economic regeneration.
4 Too little account has been taken of whether local agents in the public sector are trained to implement policies of local economic development.
5 There has been a continuing difficulty in developing coherence amongst sets of policies in different fields and across different departments.
6 The clash between central and local government has created a conflict which has limited the effectiveness of many of the most imaginative schemes.

These on-going debates surrounding both the aims and the methods of urban policy will continue to have to face the complexity and dynamism of cities. The 'good city' has been described as one in which the rungs on the ladders of opportunity are set close to each other, economically, spatially and culturally. People move most easily, for example, to a slightly more expensive house, close to their previous one. It will be easier to get that more expensive house if there are better paid jobs within reach, and easier to get the jobs if there are opportunities for a better education and further training.

This crucial interrelatedness appears to be the major challenge to effective policy intervention. In the genuinely dynamic context of cities, continually adapting the policy elements to achieve an optimal mix, and always delivering it in an effective manner, may be simply asking too much of managers of urban change.

The emphasis in the policy mix and the choice of method are largely determined by which explanation of the causes of the problem is most believed. Supporters of the concept of the cycle of deprivation lay emphasis on compensatory social work. Others believe that the problem is largely the failure of planning and administration and thus favour a more closely co-ordinated approach by bureaucrats. An approach of positive discrimination and financial incentives to achieve a redistribution of resources is viewed as being required only in the short run. Getting the underclass into employment may require very large inducements in the early stages. Alternatively greater participation by residents, and their increased control of resources, is seen by some as a method of redistributing power which will encourage a break in the cycle of decline. There are examples of all of these approaches in British attempts to tackle inner city issues. The most recent innovations – School Compacts (since 1988) and Training and Enterprise Councils (since 1990) – are local area approaches which are focusing on tackling the continuing existence of youth unemployment, ethnic unemployment and long term unemployment. They seek to improve the employability of key sections of the population, and help the underclass break out of the cycle of deprivation.

In Western capitalist societies, economic and social change will always produce urban change and with it attempts by government to ameliorate stressful social circumstances. This belief is founded on the now widespread recognition of three fundamental bases for managing urban change. First, the costs of a 'do nothing' policy are real and usually disastrous. Second, urban change has many causes and appears in many forms with great local variability. Third, public-led intervention must be truly multi-faceted and must seriously attempt to reconcile varying social and economic aims by fostering collaboration among the key local actors. In this respect inner cities have much in common with the management of economic and social change in society as a whole.

QUESTIONS

1 How has economic change contributed to inner city problems?
2 Describe the multiple-indicator approaches to measuring urban deprivation and assess their results.
3 Compare the 'physical approach' and 'social approach' to inner city policy, and explain why they have failed.
4 How far do you agree that it is not one policy, but a combination of policies, that is needed to overcome the problems of the inner city?

FURTHER READING

Robson, B. (1988) *Those Inner Cities: Reconciling Economic and Social Aims of Urban Policy*, (Clarendon Press: Oxford). A broadly based academic view of the problem and the public policy response; lengthy bibliography.

Audit Commission (1989) *Urban Regeneration and Economic Development: The Local Government Dimension*, (HMSO: London). Examines and evaluates the role of local government in supporting local business initiatives. Criticises the friction between local and central government which characterises 'the patchwork quilt of central government programmes' designed to assist urban regeneration. (A report advised by the LSE Geography Department.)

Cabinet Office, *Action for Cities* (1988), *Policy for the Inner Cities* (1977), Cmnd 6845 (HMSO: London). These represent the most recent formal statements by central government of what its policy is.

Three other useful perspectives – all in paperback:

Parkinson, M., Foley, B. and Judd, D. (1990) *Regenerating the Cities: The UK Crisis and the US Experience* (Manchester University Press: Manchester).

Faith in the City (1985) (Archbishop of Canterbury, Church House: London). Report of the Archbishop's Commission on Urban Priority Areas.

The Scarman Report: The Brixton Disorders of April 1981 (HMSO: London).

12

THE CHALLENGE OF INFORMATION TECHNOLOGY

Christopher Board

GEOGRAPHY, MAPS AND COMPUTERS

Technological change not only affects the environment and the location of economic activity, as outlined in earlier chapters, it also affects the way we do things. This chapter discusses the impact of computers on the way we study geography. Geographers have always been in the forefront of projects to use geographical information to compile atlases and maps on special themes and on countries or regions. Because geographers are primary users, they have the ability to foster understanding of the interrelationships between the geographical characteristics that make up an area.

Atlases and maps have been the traditional method of making much geographical information accessible. In the past commerce, as well as education, has had to rely mainly on a co-ordinated series of national sheet maps, referred to as a 'national cartographic system'. This system was slow, cumbersome and costly, but the technology did not exist to develop it much further. Appropriate computer software for handling spatial data had not been developed, the data themselves had not been digitised and no cheap means was available of updating the maps produced. All these problems now have technical solutions. For individuals, solutions have come to the desk top through personal computers, and for large scale commercial and public planning organisations through mini and mainframe computers.

As a result of these changes we now talk less of maps as such and more of geographical information systems that are capable of rapidly translating geographical information into maps, graphs, tabulated data or into many other formats. The impact of these changes is outlined in this chapter. First, their recent development is traced, and then their consequences are examined in respect of their application to remote sensing, teaching and commerce. Each of these fields now offers career opportunities.

NEW AND BETTER TOOLS

A pathbreaking book, *Models in Geography*, published in 1967, made much of the explosion of what was called the geographical data matrix. It was suggested that the amount of scientific information doubled every ten to fifteen years. But the editors of *Models in Geography*, Chorley and Haggett, suggested that the rate of growth of geographical information had been even more rapid than that of other scientific information. To handle this 'information explosion' they pointed to the need for an increased application of computers. They forecast that geography, because of its use of a special type of *locational* data storage, such as maps, would reap considerable benefits from the application of new electronic technology. With such advances, the problem of dealing with the massive growth of information could be managed and a greater understanding of complex global geographical systems could be achieved.

Traditionally geographers sought to manage the information explosion of locational data by making paper maps, and they created order in their worlds by making and interpreting maps. During the last two decades electronic computers have been routinely employed in all kinds of geographical analysis, but the graphical quality of their output has only slowly developed to the point where they are as good as, and sometimes better than, the traditional maps drawn by hand. These developments have been brought about by the provision of better and cheaper hardware and software mapping packages. The rapid decline in the real cost of computer processing and storage for large information sources has made computer mapping feasible in most university and college geography departments. The increasing availability of simpler, user-friendly and more flexible mapping packages has encouraged those who made maps, or have taught map design and map use, to acquire personal 'hands on' experience of computer mapping, which had hitherto been the preserve of the computer specialist. As a result, most undergraduate geography courses now include a significant amount of teaching that uses computer mapping.

The first widely available computer mapping programs were developed at Harvard University in the mid-1960s. Output was generated on standard lineprinters normally used to print text, tables and figures. An illusion of varying densities and patterns had to be created by overprinting different characters on top of each other. This left spaces around each character and produced very poor quality maps. Soon colours were introduced to replace the different combinations of characters in black. This method was used to provide maps of the 1961 and 1966 censuses for the Ministry of Housing and Local Government (now Department of the Environment). The needs of urban and regional planning were thus responsible in Britain for the first nationwide use of the national grid in computer-assisted map production. Indeed, this resulted in the first computer atlas of England and Wales.

Early in the 1970s many experiments to improve the quality of computer mapping took place. LSE's contribution, CHOROMAP, was used in the *Social Atlas of London*. Another system, GIMMS, was developed at the University of Edinburgh. This has been used by Statistics Canada for census mapping and by the Scottish Development Department. However, it was recognised that a 'macro facility' to allow users to create common groups of commands was needed. This would make the teaching of statistical mapping easier. The absence of this facility, amongst other things, discouraged the use of GIMMS for teaching computer-assisted map design.

At first, applications of the new computerised methods, such as in the *Social Atlas of London* (1974) and *People in Britain: A Census Atlas* (1980), were made possible only by major projects employing large and expensive mainframe computers. By the mid-1980s, however, off-the-shelf, user-friendly packages and the interactive use of computer hardware made small scale map design and production easier for individuals and teachers. The new packages require the user only to be computer literate, and not necessarily able to draw! More recently, smaller and cheaper automatic/ electronic scanners have made it possible to transform line images cheaply into digital records. These have greatly speeded up the preparation of outline maps, which were previously digitised manually, tracing the outline by hand with an instrument that followed each line point by point.

By 1985 the development of MAPICS at University College, London, provided a very user-friendly mapping package that completely revolutionised computer mapping in several university departments. Running on mini-computers, accessible in terms of cost to geography departments, it was readily adopted for both group teaching and research. For example, at LSE students are taught to design and produce maps interactively. This means that they apply their knowledge of the principles of map design by improving a given map in stages. Each new map is inspected and a decision made on what further improvements are necessary. All this is done on the screen. All undergraduates taking the honours geography course are now capable of producing high quality statistical maps of areas such as Greater London (Figure 12.1) to high specifications. This part of the first-year course on methods of geographical analysis provides a basis for more advanced work.

In the commercial world computers now create maps of high graphic quality, ranging from large scale digital plans to the 1 : 625,000 route planning map produced by the Ordnance Survey, as well as thousands of small scale statistical maps. Digital versions of medium scale topographical maps have been produced by several countries in Western Europe, e.g. West Germany and Norway. The 1 : 100,000 topographical series recently completed for the whole of the United States would not exist were it not for the intensive programme of digitising undertaken to make it available as a base map for the 1990 census. The production of commercial road maps and atlases is also increasingly dependent on digital data bases which are kept up to date by motoring organisations. The Automobile Association, for example, has invested in computer-assisted mapping to achieve more efficient production methods, a bigger product range and an expanding market base. Furthermore the AA has created its own data base free from any other organisation's copyright so that it can maintain it by its own updating procedures.

It is only since about 1985 that microcomputers have been able to handle satisfactorily the very large data sets with which geographers are concerned. The increasing power, speed and storage capacity of micros are such that they can now be harnessed to mapping projects which demand very large sets of locational and attribute data. The more recent IBM machines, and their clones, can cope very well with the amounts of data involved in route planning in Britain and in displaying road networks. Likewise the Apple Macintosh II and SE microcomputers are running a number of statistical mapping packages, such as MAPMAKER and MAPGRAFIX, which can drive sophisticated colour plotters.

Some examples of these advances may be familiar in school or

Figure 12.1 Computer plotting: the population density of London boroughs, 1981. The map was produced by a first year undergraduate as part of a course on methods in geographical analysis. It represents the fruit of much experimentation by trial and error and was printed on a Hewlett-Packard plotter in the graphics laboratory of the LSE Department of Geography

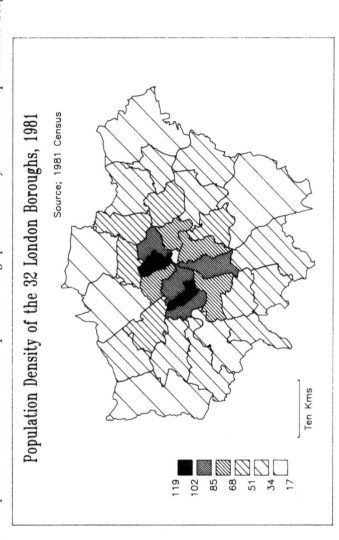

Population Density of the 32 London Boroughs, 1981

Source; 1981 Census

Ten Kms

119
102
85
68
51
34
17

college through the Domesday video discs. To commemorate the nine-hundredth anniversary of the compilation of the Domesday Book of 1086, the BBC commissioned a modern Domesday project. Storing the huge quantity of data, which are in both digital and analogue form, has been solved by new technology in the shape of two read-only memory (ROM) compact laser disks. The Domesday example shows the new power that is available to manipulate, analyse, link, display and summarise spatially referenced data by computer. Such systems are now termed geographical information systems. Although not as cheap as intended, the Domesday system is widely available in large public libraries and also in many schools and institutions of higher education.

WHAT ARE THE NEW TOOLS AND DATA USED FOR?

Geography does not change just because it has been provided with new gadgets. The new tools have, however, made the subject more effective and better able to respond to tackling new problems. This has allowed geographers, for example, to monitor environmental change and to deal with larger and more complex sets of data. Switching from scale to scale, from the global to the national, regional or local has also been made much easier.

Geographers and planners still practise their traditional tasks of describing, analysing and explaining the varied global environment with a view to its better management for the benefit of society. Statistical atlases of regions, countries and the whole world are now routinely created with computer packages. Now that data are readily available in machine-readable form, relatively inexperienced computer users can make their own sets of maps without having to resort to the draughting methods employed by the professional cartographer. Sometimes maps are even drawn automatically from information relayed from small computerised instruments in the field (data loggers) or from satellite images – no people have to intervene in production at all once the system has been set up! This is how many weather maps and environmental maps are now drawn.

It is now accepted, almost without question, that these and other modern information needs can be met only by large scale geographical information systems. As far back as 1938 the Royal Geographical Society's evidence to the Barlow Commission on the

Distribution of the Industrial Population analysed a series of maps of England and Wales, using the 'sieve' technique. This used a series of maps that excluded or included different areas possessing specific characteristics as a way of indicating where new planning developments might be avoided or located. Figure 12.2 illustrates one such exercise.

What is new about geographical information systems is the power provided by computers to analyse more data, faster and often at lower levels of aggregation than was possible hitherto. This allows 'sieving' and other methods to be tried out, criteria to be changed, and new images to be obtained, all from the desk in very rapid response times. The result is an interactive capacity to produce maps that show clearly the characteristics that it is sought to highlight. Maps and sieving, of course, are only some of the products from geographical information systems that can be used to present geographical patterns from data held in a computer system. There is also a wide range of other techniques that provoke the investigator to explore further distributions and spatial relationships through maps. One of these is remote sensing.

Remotely sensed images

The use of space platforms and satellites has very rapidly provided a further mass of data. Some of these data are photographic (e.g. from Skylab). Most are derived from sensors working in different wavebands from orbiting satellites (e.g. from the Landsat satellite). Images of cloud cover for areas ranging in size from Britain to an entire hemisphere are now routinely displayed in televised weather forecasts (derived from the Meteosat satellite). Animations displaying the development of weather systems and computer predictions of their future shape and position are also a normal part of modern weather reports. Computer print-outs of weather records all over Europe are now so commonplace and voluminous that they are used as wrapping paper once they have been examined by the forecasters at the London Weather Centre!

Less well known are the satellite images of the land surface, but even these are now becoming fashionable illustrations, chiefly because of the vast areas they include and the remarkably up-to-date detail they display. Indeed, the first atlas to be largely composed of satellite images, *The Earth from Space*, is now generally available; and it is now obligatory for virtually every new

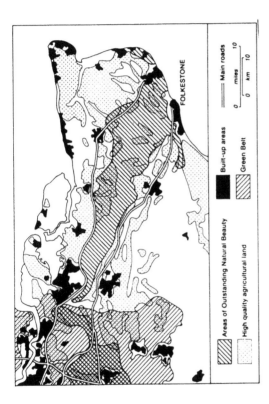

FOLKESTONE

▨ Areas of Outstanding Natural Beauty	■ Built-up areas
⬚ High quality agricultural land	▨ Green Belt

═══ Main roads

0   miles   10
0   km   10

Figure 12.2 An example of an application of the sieve technique. Sieving involves the progressive elimination of categories of land, so that areas suitable for development, without specific disqualifying characteristics, can be identified. In this example (from a Crown copyright map produced by the Department of the Environment) the restraints on development between London and the Channel tunnel are plotted on top of one another. Where possible, projected motorways or high speed rail links will avoid green belt, high quality agricultural land, Areas of Outstanding Natural Beauty, Ministry of Defence land and built-up areas. Land not affected by such restraints is conventionally left white and may therefore be considered vulnerable. In the geographical information system the boundaries of such areas are held in a data base, and any combination of zones can be produced on display to show areas free of selected restraints. Such overlays are also used to identify sites with particular qualites, e.g. for power stations or industrial complexes.

CHRISTOPHER BOARD

Plate 12.1   Remote sensing: a satellite image of the great storm over Britain in October 1987 (*Photo*: University of Dundee)

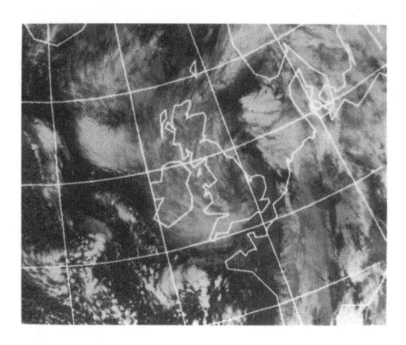

atlas to include at least one satellite image. Their chief strength lies in their ability to give a complete coverage of a large area at a single point in time, repeated at intervals of days. As a result they offer great potential to the study of very short time environmental change, such as the results of fire, flood, landslides, vulcanism, overgrazing, soil erosion or urban development.

Satellite images have also been used to assess the expansion of fast growing cities such as Algiers, Athens and Nanjing in the absence of recent conventional maps. Their use in monitoring topographical change for medium scale revision of maps in developed areas of the world is being eagerly investigated, because if they prove reliable it will save many man hours of ground survey. Satellite images can also be used as a means of assessing land use and crop yields. This proves particularly valuable in inaccessible Third World situations, where satellite images have suddenly added amazing detail on land use to what were often relatively poorly mapped regions.

250

Assessment of crop yields in the USSR has been routinely carried out by commodity brokers and NATO analysts to overcome problems of access and the inaccuracy of the USSR's own forecasts. This information has been important in the analysis of commodity market prices.

Taken together, these numerous, and generally cheap, images offer commerce, as well as students and teachers, little excuse for not appreciating the effects of geographical scale from the global to the local.

The new technology revolutionises teaching

Geographical description and analysis now thrive on the powerful arsenal of techniques that are available. A map, although produced more rapidly and in greater numbers, is still a map. Statistics can be handled and computed more efficiently. Diagrams, ranging from the familiar histograms and bar charts to more elaborate graphs, become easily produced by-products of analysis. Most under-graduates in geography now take 'methods' courses. At LSE, for example, the courses aim to develop a level of computer literacy that brings every student to the point of being able to handle statistical data and use them to design maps which do not have to be drawn by hand. Instead of the rather tedious map drafting or colouring exercises of the manual era, increasingly realistic regional case studies are used to give all students 'hands on' experience of practical computer map design. It is a comparatively simple matter to improve or to redesign a particular map, since the student spends relatively little effort in producing the final graphic image.

For geographers it is important to take appropriate decisions on designing a map so that the right effect is created. By judicious use of different output devices (plotters or printers, in monochrome or in colour) there is a vast range of possibilities that now make it feasible for any student to create his/her own statistical map to a high technical standard. However, the paradox is that it is also easier to create well produced but utterly misleading maps. This is no different from misusing a multi-coloured set of crayons. However, because maps are now so easily produced, it is easier to rush the product and have less sense of having designed the map to serve the specific purpose.

It will not surprise anyone to learn that even professional cartographers and graphic designers now use microcomputers for small scale maps and diagrams. These are destined either for desk top publishing or more conventional printing and publishing. You will have seen examples of such work in most magazines, newspapers and on television. For such maps, scanner digitisers are ideal for converting outlines of coasts and boundaries to machine-readable co-ordinates. Typically these are being pasted into 'drawing' or 'painting' packages. These allow the linework to be cleaned up, filled with patterns, annotated with text in different sizes and typefaces, or embellished with extra symbols. It takes some time to learn how to 'draw' in a freehand manner, but the ability to zoom in and enlarge sections of a map to add or remove detail is a great help. Apple Macintosh equipment in up-to-date graphics laboratories already brings this facility within the scope of any competent second or third year undergraduate. Those with some claim to artistic skill can even compose landscape sketches with the graphics packages available.

Whereas colour plotting, ink-jet and electrostatic printing have for some time enabled cartographers to produce hard copies of maps, the more recently available monitors for colour displays now open up the possibility of designing coloured statistical maps without printing them first. This will allow advanced students to design and experiment with statistical maps in colour and to examine complex policy decision-making scenarios. These types of developments are so new that the latest texts do not mention them. But they are available in the BBC's Domesday system and are becoming available in many schools through the Archimedes, Apple and Nimbus computers.

The micro-driven Domesday system contains pictorial, statistical and text data on two video disks. For example, census data for Britain can be mapped at ten different levels of spatial resolution, from 1 km squares to 10 km by 10 km squares. The operator can experiment with different sizes of square until he/she 'hits' on one which seems more meaningful. By examining the spatial correspondence of two or more variables at the same level of resolution an investigator can begin to make more sense of the human geography of Britain. Up to nine colours can be used to show which squares belong to a class in a range of a census variable, e.g. density of population, or percentage of the population over retirement age.

Wherever possible, these census data have been specially recalculated to allocate them to kilometre squares of the national grid. Unfortunately, the nine colours are arranged in the order of a so-called conventional temperature scale from cold colours to warm colours. The operator can override these colours but they always remain in the same order, so that the capacity for experimenting is extremely limited. However, it is possible to change the class limits used and thus change the pattern on the map to accommodate the new classification. These examples cover only a fraction of the information about Britain accessible on the two video discs which come with the Domesday system. But they illustrate how data in published and unpublished sources can be compactly stored and relatively rapidly accessed on a microcomputer. To carry out the same operations manually would require a large library and a staff of researchers.

The new BBC–Acorn Archimedes system, as well as other systems, which are becoming available in schools, allow the classroom to be used to develop many of these techniques. The Archimedes Draw and Paint packages allow colours to be blended from a palette into up to 256 shades displayed on the screen. Software packages for drawing and painting allow diagrams to be drawn with lines and fills of shading, all for a price of well under £1,000 per machine. These packages can all be used for mapping, either by inputing freehand through a 'mouse', or by using a scanner to input outlines from maps already available. The report of the National Curriculum Geography Working Group in 1990 responded to these new skill demands through a new curriculum. This stresses mapping and other geographical skills concerned with information technology, applied to the knowledge of places and regions. University departments, and the commercial world beyond education, will make the use of these map and graphic images a standard part of their requirements in the future. This is changing the whole environment of 'keyboard skills' that can be developed from simple text processing and computer games to a new world of graphic multi-coloured images which can reflect the imagination of the operator. It is now possible to anticipate that within a few years the techniques of geography and mapping, at a very high standard of technical sophistication, will be available in almost every school classroom, office and home.

Commercial applications: we are all geographers now!

The last ten years have seen what Chorley and Haggett forecast – serious problems of data overload. To a degree this situation has been exacerbated by our ability to collect, publish and handle increasing volumes of data. During the 1980s it became apparent that there were many parallel, duplicated and piecemeal attempts to tackle this information problem. As a result the government set up a committee of inquiry under Lord Chorley to explore how the needs of public and private users of geographical information systems might best be met. Lord Chorley's committee defined a geographical information system as 'a system for capturing, storing, checking, integrating, manipulating, analysing and displaying data which are spatially referenced to the Earth. This is normally considered to involve a spatially referenced computer database and appropriate applications software.' This has become the standard definition of a geographical information system.

Ideally a geographical information system should be able to handle data collected by complete censuses, sample surveys of households referred to postal addresses, political behaviour by ward, surveys of movement, flows along routeways and many other data sets. At a smaller scale, data from remote sensing satellites or air photography can often be fitted to mapped data and to enumerated data from censuses to provide the basis for an analysis of relationships. By recording the extent of the coincidence of selected phenomena, system users can assess the degree to which patterns are correlated. Displays of related data are a standard feature of a geographical information system. Manual cartography suffers from a history of continuous changes to the shapes and sizes of the spatial units used for producing statistics. A geographical information system can be designed to reduce the effects of such changes (although they are never completely eliminated). A geographical information system also usually incorporates some method of searching specified zones in terms of the number of areas surrounding, or at a set distance from, a point. For example, it is usually considered a requirement of a geographical information system that it should be able to relate data for areas (e.g. land use) to data from linear features (e.g. rivers and roads) or to data from points (e.g. factories or farm buildings).

The value of a geographical information system is this ability to bring the skills of the geographer to education, business and the home increasingly cheaply. This revolution in access to sophisticated geographical expertise, built into the software, is creating a new world in which everyone can have the potential to execute advanced mapping and geographical analyses.

The chief applications that have been developed in the commercial world have so far tended to focus on three areas. First, the use of geographical information systems to help maximise market potential in siting new retail developments. Here the requirement is to find the best site for a shop so as to attract most customers of the right type. The methods used are to employ a geographical information system with small scale data on household income and consumption behaviour.

The second major field of application has been the utility

Figure 12.3   The potential benefits of geographical information systems. Benefits identified by different administrative departments of Swansea City Council. All departments recognise some advantages stemmming from a single geographical information system for the city's administration.

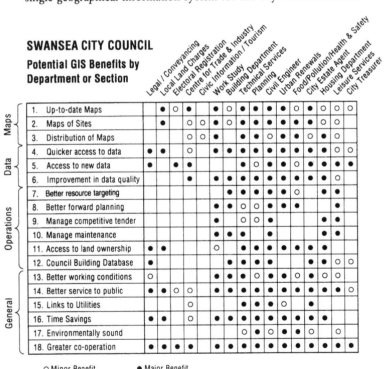

O Minor Benefit   ● Major Benefit

industries (electricity, gas, water, sewerage, telephones, cable tele-vision). The requirements of legislation, such as the Public Utilities Streets Works Act 1950, compel utilities to exchange information about their mains network and plant. Great demands have come from these industries for geographical information systems to develop large scale mapping to acceptable standards. This is used for more effective planning of resources.

The third important area of application has been in local government and other authorities. There can hardly be a single local authority now that is not thinking about how geographical infor-mation systems might help it to function more effectively by integrating, co-ordinating and pooling spatial information. This applies to such diverse areas as street lighting and street maintenance schedules, refuse collection routes, council house management, and poll tax collection. An example of this is shown for Swansea in Figure 12.3. Another example is the use by English Heritage of an automated system for dealing with scheduled ancient monuments. This information system includes digital maps to support its inventory of monuments, related to text descriptions, all of which can be easily accessible for further investigation. In other examples, air photographs and satellite imagery have played a large part in setting up a land information system for recording land use in cities and for monitoring landscape change in the ten National Parks (such as Snowdonia, the Peak District and Dartmoor).

THE FUTURE

The development of new information systems is also leading to a revolution in the requirements for how 'traditional' maps are prepared. Until very recently the Ordnance Survey's traditional large scale maps were accepted as the main data base for all geographical data in Britain. The national coverage of 1 : 10,000 maps with the national grid was published on paper or plastic and was generally not available in any other form. Responding gradually to the requirements of users who wanted up-to-date data, often in machine-readable form, the Ordnance Survey has begun to convert its large scale maps to digital format, adding the results of revision whenever possible, so that by the end of 1991 all major towns and cities will be covered by digital data. Mapping of mainly urban areas is now stored in computer systems and can be plotted to the specific requirements of customers. In this way single sheet maps can be produced (at a price!) of areas lying on four adjacent sheets, centred on, say, a factory or school. Unwanted detail can be excluded and unconventional colours can be used.

In the future the new demands for information system development is likely to come mainly from environmental requirements – reflecting the need to monitor and restrict pollution, to conserve land of high quality for agriculture, to preserve landscape as a scenic resource, to record land under various planning controls (such as green belt), or to find the best sites for development that nobody wants in their own back yard. Future geographical changes may thus be monitored by integrating information already on published maps, adding detail captured by air photography or satellite imagery and including data collected by censuses and surveys. In the near future, electronic route pricing may be established in large cities. There is already geographical system information on traffic jams. One widespread experiment, developed in 1989/90, was the installation of a geographical information system (Action Plan) in more than 150 Shell Traveller's Check shops at petrol service stations. This system displays maps, air photographs and video pictures at up to eleven map levels and scales which are downloaded from a laservision disk. By following simple instructions, merely touching the screen, the motorist or navigator can instantly obtain information about roads and traffic. Such systems are intuitive and easy to use. They give an indication of how geographical techniques are being increasingly used as part of people's daily lives.

A future without maps can hardly be imagined. But considerable research will have to address how we make better sense of maps, and how to communicate better by using maps. This will depend crucially on how we design and operate technologically sophisticated geographical information systems.

Information technology has given us access to easy and flexible approaches to maps using vastly increased quantities of data, with the potential for rapid updating. However, in order to maximise the benefits of this new technology, several obstacles will have to be overcome. As predicted by Lord Chorley's report, among the most important is access to data, often collected at public expense, but often withheld either for reasons of personal privacy or in response to political pressure. Further development will also be restricted by the sheer cost of collecting these data.

It is argued by some comentators that we still have a long way to go. 'Current geographical information systems use mainly 1980s computer technology to operationalise 1960s methods: but they do work.' In future there are many challenges in responding to the new technologies available and in anticipating the needs of users for information systems. Possible developments are the use of computer movies to explore the time dimension and the use of 'artificial intelligence' to produce new tools for geographical information systems. Both the questions 'Is there any pattern in the map?' and 'Are there significant relationships between patterns?' require a level of prior knowledge. But at present most geographical information systems have no such knowledge built into them. This represents an opportunity for geographical information systems to be developed using expert guidance systems deriving from the methods of 'artificial intelligence'.

Geographers and cartographers will be contributing to these innovations in order to develop the map as a 'friendly front end', a gateway to the data in a geographical information system. The map is thus seen as 'an initial menu of what can be viewed, rather than as an artfully coded statement of what is.'

Note: The reader should be aware that photocopying and digitising Ordnance Survey maps is subject to Crown copyright.

QUESTIONS

1 How have geographers traditionally sought to present their chief kinds of geographical information?
2 In what situations is remote sensing particularly useful?
3 What are 'geographical information systems' and what are their main areas of application?
4 What is 'interactive capacity' and how can it be applied to the 'sieving' of geographical information on computers?
5 Develop an example of how a geographical information system could be used in any project that you may have undertaken.

FURTHER READING

Maguire, D. J. (1989) *Computers in Geography* (Longman: Harlow). A valuable introduction which pulls together a wide variety of material. Useful glossary and bibliography.
Carter, J. R. (1984) *Computer Mapping Progress in the '80s* (Association of American Geographers: Washington D.C.). Introductory approach to a rapidly changing topic, but has a chapter on the microcomputer.

Boardman, D. (ed.) (1986) *Handbook for Geography Teachers* (Geographical Assocation: Sheffield), Chapters 5 and 6. On microcomputers in the classroom, maps and mapwork. Designed for teachers and written by experts.
Department of the Environment (1987) *Handling Geographic Information*. Report of the Committee of Enquiry chaired by Lord Chorley (sometimes called the Chorley report) (HMSO: London). Designed to provide guidance on how geographical data are being and should be handled in the context of a rapidly changing context of information technology. Valuable glossary and technical appendices, e.g. on the postcode system.
Rhind, D. W. (1981) 'Geographical information systems in Britain', in Wrigley, N. and Bennett, R. J. *Quantitative Geography* (Routledge & Kegan Paul: London). A brief and readable summary of the issues, policy problems and technical solutions. Includes useful examples of encoding Ordnance Survey maps.

INDEX